VIETNAM PRIMATE CONSERVATION STATUS REVIEW 2002 - PART 2

LEAF MONKEYS

TILO NADLER - FRANK MOMBERG - NGUYEN XUAN DANG - NICOLAS LORMEE

FRANKFURT ZOOLOGICAL SOCIETY - CUC PHUONG NATIONAL PARK CONSERVATION PROGRAM
FAUNA & FLORA INTERNATIONAL, VIETNAM PROGRAM
HANOI, 2003

SUPPORTED BY

Canadian Embassy Hanoi

CONSERVATION INTERNATIONAL

German Embassy Hanoi

DPZ German Primate Center

WALT DISNEY FOUNDATION

PUBLISHED BY FAUNA & FLORA INTERNATIONAL ASIA PACIFIC PROGRAMME OFFICE, IPO BOX 78, 55 TO HIEN THANH, HANOI, VIETNAM

CITATION: NADLER, T., MOMBERG, F., NGUYEN XUAN DANG & LORMÉE, N. (2003): VIETNAM PRIMATE CONSERVATION STATUS REVIEW 2002. PART 2: LEAF MONKEYS. FAUNA & FLORA INTERNATIONAL-VIETNAM PROGRAM AND FRANKFURT ZOOLOGICAL SOCIETY, HANOI, VIETNAM.

PRODUCED BY: FAUNA & FLORA INTERNATIONAL, INDOCHINA PROGRAMME OFFICE

DESIGN BY: ARIF HASYIM

COVER PHOTOS:

	3		5	
2		4		6
		1		

1 - GREY-SHANKED DOUC LANGUR, AD. MALE. *T. NADLER*
2 - CAT BA LANGUR, AD. FEMALE . *T. NADLER*
3 - BLACK-SHANKED DOUC LANGUR, AD. MALE. *T. NADLER*

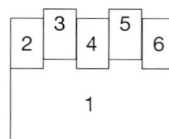

4 - DELACOUR'S LANGUR, AD. MALE. *T. NADLER*
5 - TONKIN SNUB-NOSED MONKEY, INFANT MALE, ABOUT ONE AND A HALF YEAR OLD. *T. NADLER*
6 - HATINH LANGUR, INFANT FEMALE, FOUR WEEKS OLD. *T. NADLER*

BACK COVER:
GREY LANGUR, INFANT FEMALE, ONE AND A HALF YEAR OLD. *T. NADLER*

AVAILABLE FROM:

Fauna & Flora International,
Great Eastern House, Tenison Road,
Cambridge, CB1 2TT, UK
Tel: +44 (0) 1223 571000, Fax: +44 (0) 1223 461481
E-mail: info@fauna-flora.org,
Website: www.fauna-flora.org

Fauna & Flora International
Asia Pacific Programmes
IPO Box 78, 55 To Hien Thanh, Hanoi, VIETNAM
Tel: +84 (0)4 9784470/1, Fax: +84 (0)4 9784440
E-mail: vietnam@ffi.org.vn
Website: www.flora-fauna.org

Frankfurt Zoological Society
Alfred-Brehm-Platz 16
60316 Frankfurt, Germany
Tel: +49 (0) 69 94 34 46-0, Fax: +49 (0) 69 43 93 48
E-mail: info@zgf.de
Website: www.zgf.de

Endangered Primate Rescue Center
Cuc Phuong National Park
Nho Quan District, Ninh Binh Province, Vietnam
Tel: +84 (0) 30848002, Fax: +84 (0) 30848008
E-mail: t.nadler@mail.hut.edu.vn
Website: www.primatecenter.org

Contents

Acknowledgement

Fauna & Flora International and Frankfurt Zoological Society wishes to thank the Forest Protection Department, Ministry of Agriculture and Rural Development for their active support and encouragement.

The project was funded by the Walt Disney Foundation, Margot Marsh Foundation, and the Canada Fund. We gratefully acknowledge the active support and encouragement of the Canadian Embassy.

We are grateful to Conservation International, the Embassy of the Federal Republic of Germany, Hanoi and the German Primate Center, Goettingen for support of the printing.

The field surveys for the purposes of this work were facilitated with the help and participation of the following institutions: Provincial Forest Protection Departments of Bac Kan, Binh Phuoc, Binh Thuan, Cao Bang, Dak Lak, Ha Giang, Ha Nam, Ha Tay, Hoa Binh, Kontum, Lao Cai, Nhe An, Ninh Binh, Ninh Thuan, Phu Tho, Quang Nam, Quang Ngai, Son La, Thai Nguyen, Thang Hoa, Thua Thien - Hue, Tuyen Quang, and Yen Bai, Forest Inventory and Planning Institute (FIPI), Hanoi University, Institute of Ecology and Biological Resources (IEBR), Xuan Mai Forestry College.

For assisstance and support we would like to thank: Dao Van Khuong (Director Cuc Phuong National Park), Truong Quang Bich (Vice-Director Cuc Phuong National Park), Nguyen Phien Ngung (Director Cat Ba National Park).

We would like to thank the following participants of FFI-Indochina Programme primate field surveys: Nguyen Xuan Dang (IEBR), Barney Long (FFI), Le Khac Quyet (FFI), La Quang Trung (FFI), Trinh Dinh Hoang (FFI), Le Trong Dat (FFI/Cuc Phuong National Park), Luong Van Hao (FFI/Cuc Phuong National Park), Lucy Tallents (FFI), Dong Thanh Hai (FCXM), Phung Van Khoa (FCXM), Ngo Van Tri (FFI/SIERES), and Trinh Viet Cuong (IEBR).

The Frankfurt Zoological Society would like to thank the following participants of the surveys: Lynne R. Baker (FZS), Ha Thang Long (FZS/Cuc Phuong National Park), Kieu Manh Huong (FCXM/FZS), Le Thien Duc (FZS/Cuc Phuong National Park) Luong Van Hao (FZS/Cuc Phuong National Park) and Nguyen Huu Hien FCXM/FZS).

We would like to express our gratitude to the following contributors for supplying information for this status review: Lynne R. Baker, Nicola Beharrel (Frontier Vietnam), William Bleisch (FFI), Ramesh Boonratana (aka Zimbo), Bui Huu Manh (WWF-Cat Tien Conservation Project), Bert Covert (University of Colorado), Dao Yinglan, Do Tuoc (FIPI), Dong Thanh Hai (FCXM), Roland Eve (WWF-Vu Quang Conservation Project), Neil Furey (Frontier Vietnam), Bettina Grieser-Johns, Andrew Grieser-Johns, Joanne Harding (Australian National University), Benjamin Hayes (WWF-Cat Tien Conservation Project), Huynh Van Keo (Bach Ma National Park), Le Trong Trai (FIPI/Birdlife), Li Zhaoyuan (University of Edinburgh), Barney Long (FFI-Indochina Programme), Bettina Martin (ZSCSP/Allwetterzoo Muenster), Ngo Van Tri (FFI/SIERES), Nguyen Truong Son (IEBR), Nong The Dien (Ba Be National Park), Thomas Osborn (Frontier Vietnam), Pham Nhat (FCXM), Gert Polet (WWF-Cat Tien Conservation Project), Jean-Marc Pons (Museum d'Histoire Naturelle, Paris), Elizabeth Rogers (University of Edinburgh), Christian Roos (German Primate Center), Rosi Stenke (ZSCSP/Allwetterzoo Muenster), Ulrike Streicher (EPRC), Andrew Tordoff (BirdLife International, Vietnam Programme), Trinh Viet Cuong (IEBR), Vu Ngoc Thanh (National University Vietnam, Hanoi), Vern Weitzel (UNDP, Vietnam Programme) and Yang Chang Man (Raffles Museum, Singapore).

For editing and proof reading various versions of this report we would like to thank Catherine Banks, Bert Covert, David Chivers, Barney Long, Richard Rastall, and Ulrike Streicher.

Grateful thanks are also owed to the many local people who helped the survey teams in the field.

Conventions

The status and distribution of each species of leaf monkey is described in detail for the whole of Vietnam. In addition, to place the species in a regional context, a brief overview is given of distribution and status in other areas of Indochina. In this work, we consider Indochina (as a zoogeographic region) to comprise the following regions: the Socialist Republic of Vietnam, the Lao People's Democratic Republic, the Kingdom of Cambodia and southern parts of Yunnan and Guangxi provinces in the People's Republic of China.

For each species, a list of records in Vietnam is given. These lists are fairly extensive given that they include a number of unpublished scientific records.

The occurrence of the species at a given locality is considered as "confirmed" only if direct evidence (sighting, vocalization) has been recorded since 1989 by a reliable collector. It is considered as "provisional" if the species was recorded by indirect evidence (specimen, report by local people). If no records more recent than 1988 could be found during our study, the status of the species at the considered locality is considered "unknown".

Specimens reported from provincial capitals are not taken into account in this work. Such localities are often the most important human settlement in the region, and, therefore, it is not possible to accurately ascertain the exact origin of the specimen, as it could have been collected from a number of areas in the province. Therefore, we consider that, if a specimen was reported from such a locality, its provenance cannot be determined with any certainty.

Forest types given in Appendix 2 follow the classification used in Wege *et al.* (1999). Seven types are considered in this work: evergreen, coniferous, semi-deciduous, deciduous, bamboo, mixed and limestone. Evergreen forest is itself divided into lowland and montane evergreen according to altitude, with areas below 700 m referred to as lowland, and above 700 m as montane.

"Special-use forest" is referred to in the text; this describes only protected areas: national parks, nature reserves and cultural and historical sites. This classification is based on a classification with a legal basis founded in 1991. Protected areas, both with governmental decree and provincial decree were established on this basis. A new classification of protected areas was given legal status in 2001. The existing protected areas, however, have not been correspondingly officialised.

The size of Special-use forests refers only to the Sourcebook of Existing and Proposed Protected Areas in Vietnam (Birdlife International and the Forest Inventory and Planning Institute, 2001). But because boundaries are often revised following the original decree and sometimes are not correctly established, a degree of uncertainty exists over the precise size of many protected areas.

Abbreviations

Key to abbreviations for museum collections:

AMNH	American Museum of Natural History, New York
BMNH	British Museum (Natural History), London
FCXM	Forestry College of Vietnam, Xuan Mai
FMNH	Field Museum of Natural History, Chicago
IEBR	Institute of Ecology and Biological Resources, Hanoi
IZCAS	Institute of Zoology, Chinese Academy of Sciences, Beijing
KIZ	Kunming Institute of Zoology, Kunming, China
MNHN	Muséum National d'Histoire Naturelle, Paris
USNM	United States National Museum of Natural History, Washington D.C.
ZMVNU	Zoological Museum, Vietnam National University, Hanoi
ZRC	Zoological Reference Collection, Department of Zoology, National University of Singapore

Key to other abbreviations used in this work:

a.s.l.	above sea level
BirdLife	BirdLife International
CPAWM	Center for Protected Areas and Watershed Management of department of Forestry, Ministry of Agriculture and Forestry (Lao PDR)
EPRC	Endangered Primate Rescue Center, Vietnam
FFI	Fauna and Flora International
FIPI	Forest Inventory and Planning Institute
FPD	Forest Protection Department
FZS	Frankfurt Zoological Society
IEBR	Institute of Ecology and Biological Resources
IUCN	World Conservation Union (International Union for the Conservation of Nature and Natural Resources)
Lao PDR	Lao People's Democratic Republic
MARD	Ministry of Agriculture and Rural Development
MoF	Ministry of Forestry
NBCA	National Biodiversity Conservation Area (Lao PDR)
PNBCA	Proposed National Biodiversity Conservation Area (Lao PDR)
SEE	Society for Environmental Exploration
SFNC	Social Forestry and Nature Conservation in Nghe An Province
SIERES	Institute of Tropical Biology, Sub-Institute of Ecology, Resources and Environmental Studies
UNESCO	United Nations Education Scientific and Cultural Organisation
Vietnam NCST	Vietnam National Center for Natural Science and Technology
VRTC	Vietnam-Russian Tropical Center
WCS	Wildlife Conservation Society
WWF	World Wide Fund for Nature
ZSCSP	Zoological Society for the Conservation of Species and Populations

1. Introduction

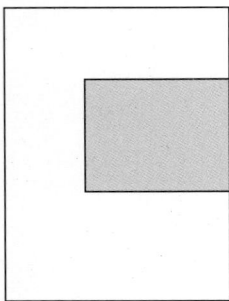

Red-shanked douc langurs; group of subadult males.

T. Nadler

1.   Introduction

1.1   Primates on the Brink of Extinction

The Primate Conservation Status Review, Vietnam 2002 contains alarming findings for the fate of Vietnam's primates. At the beginning of the new millennium, among the many species that could be lost world-wide are mankind's closest relatives, the non-human primates. In Vietnam, many primates are either endangered or critically endangered. Three of the four species that are endemic to Vietnam are critically endangered: Cat Ba or Golden-headed langur (*Trachypithecus p. poliocephalus*), Delacour's langur (*T. delacouri*) and the Tonkin snub-nosed monkey (*Rhinopithecus avunculus*) whilst the Grey-shanked douc langur (*Pygathrix cinerea*) is 'data deficient' and may well, in reality, be critically endangered. These four primates are among the most endangered primates on earth (Conservation International and IUCN Primate Specialist Group, 2000). Without rigorous and immediate conservation interventions, these species are facing global extinction. Other primates, such as Western black-crested gibbons *Nomascus concolor* and one of the world's most critically endangered primate, the Eastern black-crested gibbons *Nomascus nasutus*, are also facing extinction in Vietnam.

Amazingly, despite globally increasing rates of deforestation and hunting, there has been only one documented extinction of any primate species or sub-species on earth during the last millennium. It is a major challenge for the 21st century to save Vietnam's primates and their contribution to the world's natural heritage.

This conservation status review provides detailed recommendations for each species, ranging from research, protected area gazettement and anti-poaching patrols to conservation awareness. The most urgent conservation interventions are:

*   Implementation of on-the-ground community-based conservation and law enforcement interventions for the protection of the Tonkin snub-nosed monkey. Conservation interventions need to be undertaken to mitigate the immediate threats related to the dam construction on the Gam River adjacent to Na Hang Nature Reserve.

*   Improvement of protection for the endemic Cat Ba langur on Cat Ba Island and the development of in-situ and ex-situ population management strategies.

*   Improvement of protection and law enforcement for the protection of the Delacour's langur as well as the development of in-situ and ex-situ population management strategies.

*   Further field surveys both to assess the conservation status of the endemic Grey-shanked douc langur, most probably a "critically endangered" species and to expand the protected area network for this species.

*   Further field surveys to assess the status of the Grey langur in Vietnam and to strengthen the current protected area network in areas of this species' occurrence.

*   Further field surveys to assess the conservation status of the national most probably "critically endangered" Francois' langur, as well as an expansion of the protected area network to incorporate threatened sub-populations.

1.2   Background

Vietnam extends about 1,600 km from north to south, along the eastern coast of Indochina from 23°N to 8°30'N. The country covers about 331,689 km². Altitude varies from sea level to 3,143 m at the summit of Mount Phan Si Pan in the extreme north-west of the country. Plains are principally found in the Mekong and Red River deltas, which are linked by a narrow coastal plain. The rest of the country comprises hills and high mountain ranges.

A geological feature of major interest is the limestone karst formations that are mainly found in the central and northern parts of the country. These areas support a high degree of endemic fauna and flora. Almost all of Vietnam's endemic and critically endangered primate species live either in limestone forests or montane forests, the exception being the Grey-shanked douc langur.

In 1995, natural forest covered 8,769,000 ha, or 27.5% of the land area of Vietnam. Following the classification used by Wege *et al.* (1999), three forest types should be considered of particular importance for biodiversity. Evergreen forest, which is found in areas with a regular, high rainfall, is the main forest type in Vietnam, accounting for 64% of the total area of natural forest. Semi-deciduous and deciduous forests, which occur in lowland areas experiencing a distinct dry season, are largely restricted to central and southern parts of the country. In areas dominated by deciduous and semi-deciduous forests, leaf monkeys and gibbons are limited to evergreen forest patches along streams and on hills. A third forest type of interest is limestone forest, restricted to karst formations. Bamboo and coniferous forests, while natural, are essentially secondary in nature and their biodiversity value is lower than other natural forest types (Wege *et al.*, 1999).

8.6% of Vietnam's natural forest is now within the protected areas network. Special-use forests, including national parks, nature reserves and cultural and historical sites cover 2,370,270 ha (Nguyen Ba Thu, 2001).

1.3   Vietnam Primate Conservation Status Review

The Vietnam Primate Conservation Status Review began implementation in July 1999, with the objective of collating a comprehensive data set for leaf monkeys and gibbons in Vietnam. The results are contained within two publications. This publication covers leaf monkeys, whilst gibbons are covered by a separate publication (Geissmann *et al.*, 2000. Vietnam Primate Conservation Status Review 2000. Part 1: Gibbons).

Methods

Data on the status of leaf monkey species in Vietnam was collated from several sources:

*   Scientific reports. Most available reports about leaf monkeys, their habitats or their conservation status in Vietnam were analysed. Field survey records were considered reliable where they were given by a reliable surveyor, and the source of the information and the exact date and location were given. However, given the scarcity of information, all data detailing the source and its level of evidence were included in this work.

*   Museum specimens. The origin of each museum specimen of Vietnamese leaf monkeys was collated from the museum itself or from scientific literature. The exact location or the method of collection are often unknown and, therefore, these data should be considered as provisional. However, this type of data provides invaluable information regarding the historical distribution of given species, particularly in areas where the animals no longer occur.

✱ Personal communication. A significant quantity of information concerning primates is still unpublished in Vietnam. The authors appealed to numerous people, who for one reason or other were working in the field, to record any opportunistic sightings or other provisional information regarding primates.

✱ FFI primate field surveys. From October 1999 to March 2003, numerous primate field surveys were carried out in northern Vietnam. Because the objective was to collect data on species' status and distribution in remote areas, only a short time was spent at each locality and as large an area as possible was covered. The leaf monkey surveys were particularly focused on the Tonkin snub-nosed monkey *Rhinopithecus avunculus* and the Francois langur *Trachypithecus francoisi*. The list of field surveys, with the fieldworkers involved and the species targeted, are given in the Table 1.3-1.

✱ Results of field surveys conducted by other organizations and institutions than FFI for the Vietnam Primate Conservation Status Review. The list of these main field surveys are given in Table 1.3-2.

Table 1.3-1

Field surveys conducted by FFI for the Vietnam Primate Conservation Status Review - Leaf Monkeys

Species targeted	Location of the field survey	Date of survey	Participants
Tonkin snub-nosed monkey	Bac Kan, Thai Nguyen, Tuyen Quang Prov. Ha Giang Prov.	Oct.-Nov. 1999 Dec. 2000-Jan. 2001	Dang Ngoc Can, Nguyen Truong Son (IEBR) Le Khac Quyet, Barney Long (FFI)
Francois' langur	Bac Kan Prov.	Feb.-March 2000	Ngo Van Tri (FFI/SIERES), Nicolas Lormee (FFI)
	Ha Giang Prov. Cao Bang Prov.	Dec. 2000-Jan. 2001 Feb.-March 2001	Le Khac Quyet (FFI) Le Khac Quyet, La Quang Trung (FFI)
	Cao Bang Prov. Cao Bang Prov. Kim Hy NR- Bac Kan Prov.	April-May 2001 April-May 2001 June 2001	Trinh Dinh Hoang (FFI) La Quang Trung (FFI) La Quang Trung, Trinh Dinh Hoang (FFI)
	Bac Kan Prov.	Feb-March 2003	Le Khac Quyet (FFI) Trinh Dinh Hoang (FFI) Gabriella M.Fredriksson (FFI)
Phayre's langur	Yen Bai Prov.	April 2001	Andrew Tordoff (BirdLife), Le Trong Dat (FPD/FFI), James Hardcastle (FFI)
	Thanh Hoa Prov. Nghe An Prov.	April 2002	La Quang Trung (FFI) Trinh Dinh Hoang (FFI)

Table 1.3-2
Field surveys conducted by other organizations and institutions than FFI for the
Vietnam Primate Conservation Status Review - Leaf Monkeys

Species targeted	Location of the field survey	Date of survey	Participants
Delacour's langur	Thanh Hoa Prov.	Mar.-Apr. 1999	Lynne Baker, Luong Van Hao, Tilo Nadler (FZS)
	Ninh Binh Prov.	July-Sept. 1999	Luong Van Hao, Tilo Nadler (FZS)
	Hoa Binh Prov.	Jun.-July 1999	Luong Van Hao (FZS)
	Ha Tay, Son La, Ninh Binh Prov.	Jun.-Oct. 1999	Lynne Baker, Ha Thang Long, Tilo Nadler (FZS)
	Thanh Hoa Prov.	Jun. 1999	Lynne Baker, Ha Thang Long (FZS)
	Hoa Binh, Thanh Hoa Prov.	Jan. 2000	Luong Van Hao (FZS)
	Hoa Binh Prov.	Mar.-Apr. 2000	Luong Van Hao (FZS)
	Thanh Hoa, Nghe An Prov.	Aug. 2000	Luong Van Hao (FZS)
	Cuc Phuong NP- Ninh Binh Prov.	Sept.-Dec. 2000	Luong Van Hao, Tilo Nadler (FZS)
	Van Long NR, Ninh Binh Prov.	Mar.-Apr. 2001	Nguyen Huu Hien, Kieu Manh Huong (FCXM/FZS)
Cat Ba langur	Cat Ba Island	Jan. 1999	Lynne Baker (FZS)
	Cat Ba Island	Nov. 1999-Mar. 2000	Tilo Nadler, Ha Thang Long, Marco Mehner (FZS)
	Cat Ba Island	2001	Rosi Stenke (ZSCSP)
Douc langur	Quang Ngai, Binh Dinh Prov.	May 2000	Ha Thang Long (FZS)
	Thua Thien Hue, Kon Tum, Quang Ngai, Quang Nam Prov.	May-July 2000	Ha Thang Long (FZS)
	Quang Nam Prov.	Oct. 2000	Ha Thang Long (FZS)
	Phong Nha-Ke Bang, Quang Binh Prov.	Oct.-Dec. 2000	Pham Nhat (FCXM), Nguyen Xuan Dang (IEBR) et al.
	Ninh Thuan, Binh Thuan Prov.	Aug. 2001	Ha Thang Long (FZS)
	Dak Lak, Binh Phuoc Prov.	Nov.-Dec. 2001	Ha Thang Long, Le Thien Duc (FZS)

2. Leaf Monkeys

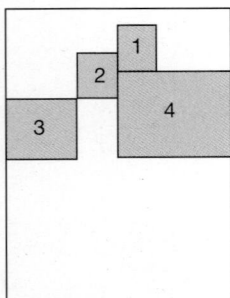

1. Indochinese Silvered langur; ad. female. *T. Nadler*

2. Hatinh langur; infant female, four weeks old. *T. Nadler*

3. Grey langur; ad. male. *T. Nadler*

4. Red-shanked douc langur; ad. male. *T. Nadler*

2. Leaf Monkeys

2.1 An introduction to the leaf monkeys

The order Primates includes about 350 species and around 630 taxa (monotypic species and subspecies). Two suborders are recognized: the Strepsirhini (lower primates) with 8 families and the Haplorhini (higher primates) with 7 families.

The higher primates are distributed with four families in the Neotropics (Cebidae, Nyctipithecidae, Pitheciidae, and Atelidae), with two families in Asia (Tarsiidae and Hylobatidae), two families in Africa and Asia (Cercopithecidae and Hominidae), and the Humans - also belong to the Hominidae - globally.

The Neotropic families are grouped in the Infraorder Platyrrhini (New World primates), whereas the Old World families grouped in the Infraorder Catarrhini. Cercopithecidae (Old World monkeys) include two extant subfamilies: Cercopithecinae and Colobinae. Cercopithecinae, often referred to as the cheek-pouched monkeys, include the macaques, baboons, drills, mangabeys, and guenons.

Colobinae are often referred to as leaf monkeys because of their predominantly specialized folivorous diet. In fact, many of the anatomical features distinctive to the Colobinae are related to their folivorous adaptation.

While all cercopithecoids eat a variety of foods, colobines are adapted to digest cellulose and denature toxins in leafy materials, though several species are not predominantly folivorous. This adaptation has implications for the design of the digestive system as well as for ecology and foraging behaviour.

Leaf-monkeys possess a highly specialised stomach allowing the fermentation of the cellulose, which constitutes the main part of plant material (Oates & Davies, 1994). The stomach of the Colobinae, is multi-compartmented and consists of a forestomach or saccus, the tubus gastricus and the pars pylora. The phenomenon of fermentation occurs in the forestomach. This part is enlarged and multilobed and possesses a high quantity of micro-organisms (namely bacteria but sometimes fungi) able to produce the enzymes that break-down cellulose. The forestomach also provides an environment of lower acidity than the toxic environment of the stomach. Some research has found a pH of only 5.5 to 7 in the saccus compared to a pH of 3 in the tubus gastricus (Chivers, 1994). An enlarged liver in most colobines processes the toxins, which many trees proceduce to protect mature leaves against predation. Other characteristics of the Colobinae can also be related to their diet. Molar cusps of the leaf-monkeys, in contrast to those of the Cercopithecinae, are high, and pointed, and have long shear crests and are thus better adapted to finely mince leaves for digestion and the salvary glands are enlarged to support the digestion (Oates & Davies, 1994).

Additionally, colobines lack cheek pouches, which are used to keep food for a short time. A summary of morphological features differentiating the Colobinae and the Cercopithecinae are listed in table 2.1-1 (Napier & Napier, 1985; Oates & Davies, 1994).

The Colobinae subfamily is represented by two major evolutionary radiations (*Oates et al.*, 1996; Novak 1999; Groves 2001) one in Africa and the other in Asia. Genetic studies support this classification (Forstner *et al.*, 1997; Zhang & Ryder, 1998).

The African colobines, commonly called colobus monkeys are comprised of two (*Colobus, Procolobus* Oates, 1996) or three (*Colobus, Procolobus, Piliocolobus* Groves, 2001) genera.

The Asian colobines commonly referred to as langurs are comprised of seven genera (*Semnopithecus, Trachypithecus, Presbytis, Pygathrix, Rhinopithecus, Nasalis* and *Simias*) with about 35 to 43 species and 90 to 93 taxa (Eudey, 1997; Groves, 2001).

There is no consensus for the use of the common name "langur" and common names for subgenera or species groups. Originally this Hindi word "langur" was used for the long-tailed monkeys (Oates *et al.*, 1994) but has been used by subsequent authors in both general and specific senses. Some publications have used "langur" to include every Asian Colobinae while others have limited the term to *Semnopithecus*, *Presbytis* and *Trachypithecus*. In some cases, "langur" has been used without making reference to the genus.

The following names will be used in the present work: "langur" will refer to *Trachypithecus*, "douc langur" refers to *Pygathrix* and "snub-nosed monkey" for *Rhinopithecus*.

Table 2.1-1
Morphological differences between Colobinae and Cercopithecinae

Colobinae	Cercopithecinae
Large multichambered stomach	Simple stomach
Enlarged liver	small liver
No cheek pouches	Cheek pouches
Enlarged salivary glands	small salivary glands
Nasal bones short and broad (except *Nasalis* and *Simias*)	Nasal bones long and narrow
Interorbital width wide	Interorbital width narrow
Deep jaw	Shallow jaw
Molar teeth with high pointed cusps	Molar teeth with low rounded cusps (except *Theropithecus*)
Hands and feet are long and slender	Hands and feet shorter
Thumb absent or very short	Thumb well-developed
Tail very long	Tail short or medium

2.2 Leaf Monkeys in Vietnam

Vietnam possesses a very diverse range of primates compared to other South-East Asian countries, 23 to 25 primate taxa, species and subspecies, are currently recognized. Ten taxa belong to the leaf monkeys. Three are endemic: Cat Ba langur (*Trachypithecus p. poliocephalus*), Delacour's langur (*Trachypithecus delacouri*) and Tonkin snub-nosed monkey (*Rhinopithecus avunculus*). The recently described Grey-shanked douc langur (*Pygathrix cinerea*) at this time has not been recorded in another country. The list of leaf monkey species and sub-species known to inhabit Vietnam are given in the Table 2.3-1.

Despite this exceptional diversity and the international interest which it arouses, there is still very little scientific data available on the distribution and status of primates in the country. This can be explained by the long isolation of the country due to several decades of war and political ostracism. In recent years, a number of surveys have been conducted and our knowledge has considerably increased. However, at this time no comprehensive work has been published which compiles all of the data on this subject from the whole country.

This lack of knowledge about the status and distribution of leaf monkeys in Vietnam poses a serious problem in terms of how to conduct a long-term conservation programme. The fact is that the recent surveys carried out in Vietnam reveal some alarming statistics. Of particular note is the current plight of the critically endangered Tonkin snub-nosed monkey, Delacour's langur and the Cat Ba langur.

Some primate surveys, however, have been conducted in Vietnam in the past. In recent times, Ratajszczak *et al.* (1990) gave a preliminary idea of the distribution and status of gibbons and langurs in northern Vietnam. Data from southern Vietnam was published by Eames & Robson (1993). The main field surveys publications for leaf monkeys in Vietnam prior to the new data published in this work, are listed in table 2.2-1.

Table 2.2-1

Main field survey publications for leaf-monkeys in Vietnam

Targeted species	Reference
Leaf-monkeys and gibbons in northern Vietnam	Ratajszczak *et al.*, 1990
Tonkin snub-nosed monkey	Ratajszczak *et al.*, 1992; Boonratana & Le Xuan Canh, 1994
Douc langur and gibbons in southern Vietnam	Eames & Robson, 1993
Douc langur	Lippold 1995a, 1995b; Pham Nhat, Do Quy Huy & Pham Hong Nguyen , 2000
Hatinh langur	Pham Nhat *et al.*, 1996b
Cat Ba langur	Nadler & Ha Thang Long, 2000
Delacour's langur	Nadler & Ha Thang Long, 2001

2.3 Molecular phylogeny and systematics of Vietnamese leaf monkeys (Christian Roos)

The phylogenetic relationships among different Colobinae taxa are still controversially discussed due to the lack of available material. The currently accepted classification of langurs is mainly based on morphological studies, while genetic data still play only a subordinate role. However, in recent time, new, highly sensitive technologies have come up in the field of molecular genetics, which permit analysis of even less DNA-carrying materials, such as faeces or single hairs.

As part of a study on the complete Colobinae evolution, the Vietnamese leaf monkeys play a key role, because of the high number of endemic species and the low number of currently available data.

In order to solve the phylogenetic relationships among langurs and to clarify their systematic position, we analysed a 570 base pairs long fragment of the mitochondrial cytochrome b gene from all currently recognised Vietnamese langur taxa (Table 2.3-1), following standard methods as described in Roos & Nadler (2001). Phylogenetic tree reconstructions and pairwise distance comparisons carried out on the data set bring new insights into the evolution of these primates. The obtained tree topologies and their branching pattern are well resolved and have high statistical support (Fig. 2.3-1).

Relationship among genera

In Vietnam, three genera are recognised: *Trachypithecus*, *Rhinopithecus* and *Pygathrix*.

The data set includes four genera of Colobinae monkeys (Fig. 2.3-1). The only African genus, *Colobus* was used as outgroup for tree reconstructions.

Within the three Asian genera, *Trachypithecus* was the first to split off, followed by *Rhinopithecus* and *Pygathrix*. These results reflect previous classifications and genetic data very well, in that *Rhinopithecus* and *Pygathrix* share a common ancestor. However, the classification of both taxa into genera or only subgenera is still discussed. Based on these data, it would be justified to separate both taxa at the genus level.

Relationship within *Trachypithecus*

The genus *Trachypithecus* is the most wide-spread Asian langur genus of Asian langurs with a distribution area from India to Vietnam and Sumatra, Java and Borneo. In Vietnam, in the past, taxa of three major groups were recognized:

cristatus group

a Silvered langur closely related to other species of this group on Java, Borneo, Sumatra, some neighbouring islands, Malay peninsula; described as *T. cristatus*

obscurus group

a Grey langur resembling most closely the Phayre's langur, and hence, it was described as a subspecies of the Phayre's langur (*T. phayrei crepusculus*),

francoisi group, mostly specified as the superspecies *francoisi*

with six to seven different taxa of which four to five were recognized for Vietnam (*francoisi, poliocephalus, delacouri, hatinhensis, ebenus*).

The data set excludes four species of the genus *Trachypithecus*, since these (*johnii, vetulus, geei* and *pileatus*) are more closely related to the genus *Semnopithecus* than they are to *Trachypithecus* and hence should be recognised as species of *Semnopithecus*. These species do not occur in Indochina.

Following the results of the DNA analysis within the genus *Trachypithecus*, a first radiation into three main lineages occurred:

cristatus group; a Silvered-, Ebony langur group with three species (*germaini, cristatus* and *auratus*),

obscurus group; a Phayre's-, Dusky langur group with two species (*phayrei, obscurus*)

francoisi (superspecies) group including *crepsuculus*.

Later, the Silvered langurs radiated again and the data show that the Silvered langurs should be separated into three distinct species. Now they have the largest distribution area of all Asian langurs. One species on Java (*auratus*), one on Sumatra, Borneo and Malay peninsula (*cristatus*) and one in Thailand, Cambodia and South Vietnam (*germaini*). The monophyly of the Silvered langur group is highly supported with a bootstrap value of 95%.
We propose to use the name Indochinese Silvered langur for *Trachypithecus germaini*.
In contrast to previous assumptions, the data have shown that *crepusculus* (previous *T. phayrei crepusculus*) is not a close relative of the Phayre's langur group, but a distant relative of the superspecies *francoisi*. The monophyletic origin of the superspecies *francoisi* and *T. crepusculus* is highly supported in the phylogenetic tree with a bootstrap value of 96%.
Based on these data, it is justified to separate *crepusculus* from other taxa of the Phayre's langur group as a separate species, *Trachypithecus crepusculus*.
Based on this new systematic position there is now no common name for this taxa.
We propose to use the name Grey langur, based on the Vietnamese common name for this species.

The superspecies *francoisi* (Roos *et al.*, 2001), comprises six closely related langur taxa, which are endemic to Northern Indochina. The monophyletic origin of this group is highly supported with a bootstrap value of 99%. Within the group, three major genetic lineages were detected that reflect the geographic distribution pattern of these taxa very well:

one in the Northern distribution area containing the taxa *francoisi*, *poliocephalus* and *leucocephalus*;

one in the Southern part of its distribution area with *laotum* and *hatinhensis*, and

one taxa distributed between the Northern and Southern group, *delacouri*.

Within the Northern group, *T. francoisi* was the first to split off, with *poliocephalus* and *leucocephalus* as the last to diverge. This well reflects the differences and similarities in fur coloration and behaviour among these primates. Based on the molecular data, it would be justified to separate *T. francoisi* from the two other taxa as distinct species and to recognize *poliocephalus* and *leucocephalus* as subspecies of the species *T. poliocephalus*.

The two taxa represented in the Southern lineage are very closely related and hence should be recognised only as subspecies of the species *T. laotum*.

In recent time, a new subspecies, *T. auratus ebenus*, was described (Brandon-Jones, 1995), which resembles more closely the *francoisi* group than the Ebony langur from Java (Nadler, 1998d, 1998e). Molecular genetic studies using the type specimen (National Museum of Natural History, Smithsonian Institution, Washington) and the only living all-black langur, kept at the EPRC, have shown that this taxon cannot be separated from the Hatinh langur (*T. l. hatinhensis*) and hence, should not be recognised as a distinct subspecies. It can be expected that the Black langur is only an melanistic morph of the Hatinh langur, which is also supported by field observations in that both morphs are present in the same region (Ruggieri & Timmins, 1995).

T. delacouri, distributed between both groups is more closely related to the southern group than to the northern ones. Based on genetic distances, however, it is justified to separate this taxon also as a distinct species.

Relationship within *Rhinopithecus*

Within the genus *Rhinopithecus*, four different species are recognized. Three of them, *brelichi*, *bieti* and *roxellana* are endemic to China, whereas the fourth species, *R. avunculus* can only be found in Northern Vietnam. Three of these species were part of the study. Based on the genetic distances between analysed taxa, it is justified to separate them all as distinct species. However, the relationship among *bieti*, *roxellana* and *avunculus* is not very well resolved. Although *avunculus* represents the sister clade to *bieti* and *roxellana* in the tree, these results should be treated with caution, since the statistical support for this relationship is very low (52%), and could change with a higher number of sequences and specimens.

Relationship within *Pygathrix*

The douc langurs, genus *Pygathrix*, are endemic to Indochina. Previously, two taxa, a red-shanked (*nemaeus*) and a black-shanked form (*nigripes*) were recognized, which most authors have included within one species *P. nemaeus*. However, Nadler (1997) described a new form with grey shanks as *P. nemaeus cinereus*. Later, the name was altered with grammatical reason to *P. nemaeus cinerea* (Brandon-Jones in: Timmins & Duckworth, 1999). Another taxa, *P. nemaeus moi* from the Langbian Peak (Lam Dong Province, Vietnam) was already described in 1926 by Kloss, which most scientists have synonymized with the black-shanked form, *P. nemaeus nigripes*.

Based on genetic distance comparisons, *P. nemaeus moi* cannot be separated from the Black-shanked douc. These results confirm previous assumptions and hence, we follow these studies and synonymize *moi* with *nigripes*.

The genetic distances between the other three taxa are within the range typical for those between species, which has led to the conclusion to separate all three taxa as distinct species (Roos & Nadler, 2001). Phylogenetic tree reconstructions carried out on the data set resolve the relationship among the three taxa with high support. *P. nigripes* branched off first, followed by *cinerea* and *nemaeus*. These relationships fit well with the distribution pattern observed in douc langurs.

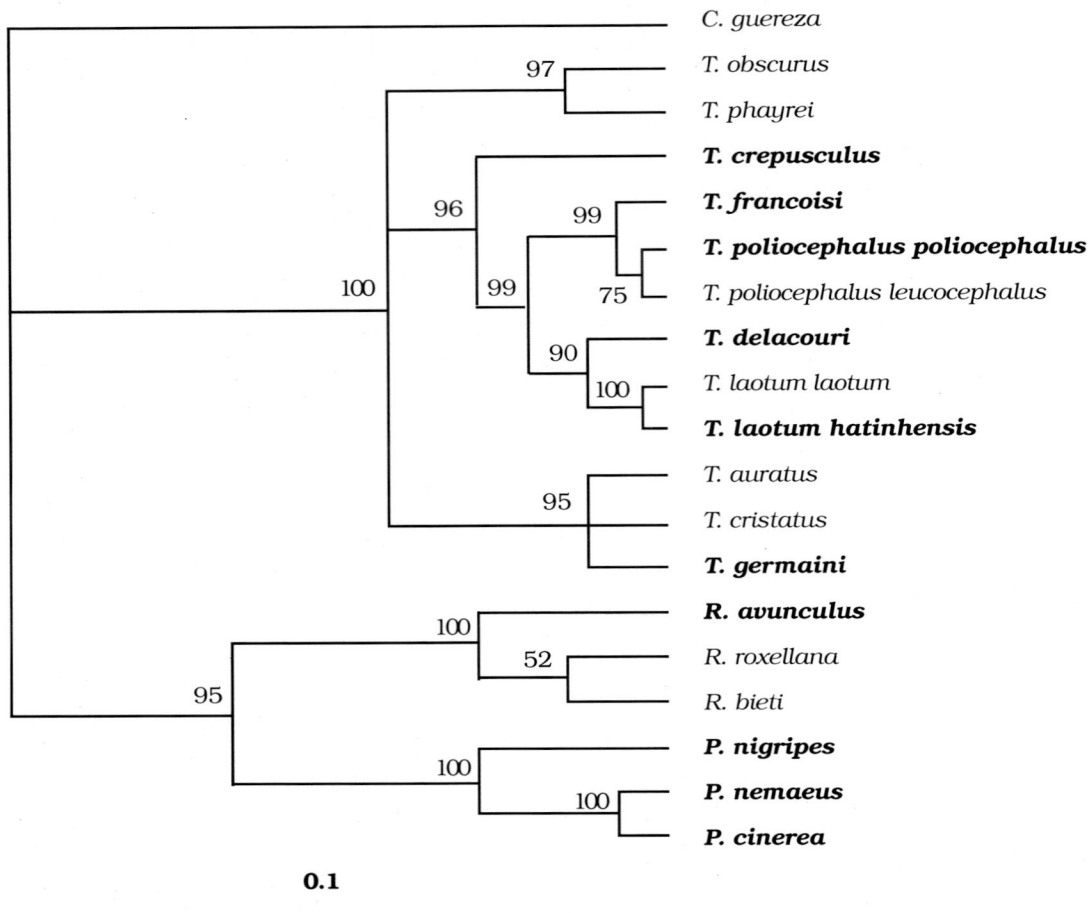

Fig. 2.3-1.
50%-majority-rule consensus tree for the maximum-likelihood method following a molecular clock. Branch lengths are drawn according to the number of substitutions per position with the bar indicating 0.1 substitutions per site. Numbers in the tree indicate bootstrap values in %. Species names in bold represent Vietnamese primates.

Table 2.3-1
List of the leaf monkey species and subspecies known to occur in Vietnam

Common name	Scientific name Genus	Species	Subspecies	
Francois' langur	*Trachypithecus*	*francoisi*	-	
Hatinh langur	*Trachypithecus*	*laotum*	*hatinhensis*	①
Cat Ba langur	*Trachypithecus*	*poliocephalus*	*poliocephalus*	
Delacour's langur	*Trachypithecus*	*delacouri*	-	
Grey langur	*Trachypithecus*	*crepusculus*	-	②
Indochinese Silvered langur	*Trachypithecus*	*germaini*	-	③
Red-shanked douc langur	*Pygathrix*	*nemaeus*	-	
Black-shanked douc langur	*Pygathrix*	*nigripes*	-	
Grey-shanked douc langur	*Pygathrix*	*cinerea*	-	
Tonkin snub-nosed monkey	*Rhinopithecus*	*avunvulus*	-	

① Including the black morph of this subspecies, described as *Trachypithecus auratus ebenus* and subsequently placed on different systematic positions
② Formerly treated as Phayre's langur *Trachypithecus phayrei crepusculus*
③ Formerly treated as Silvered langur *Trachypithecus cristatus*.

3. Langurs

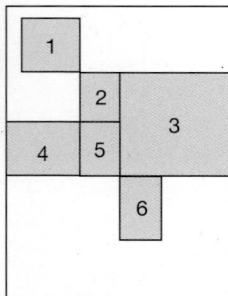

1. Delacour's langur; subad. male. *T. Nadler*

2. Francois' langur; ad. male. *T. Nadler*

3. Cat Ba langur; juv. female, about one and a half year old. *T. Nadler*

4. Grey langur; juv. female, one and a half year old. *T. Nadler*

5. Francois' langur; ad. female. *T. Nadler*

6. Indochinese Silvered langur; ad. female. *T. Nadler*

3.   Langurs

3.0   An introduction to the langurs (genus: *Trachypithecus*)

3.0.1   Taxonomy

The genus *Trachypithecus* Reichenbach, 1862 includes 15 to 17 species of leaf-monkeys of approximately the same size.

Similarities in morphology have prompted some taxonomists (Brandon-Jones, 1984, 1995; Corbet & Hill, 1992), to sink *Trachypithecus* into the genus *Semnopithecus* Desmarest, 1822, citing evidence of hybridisation between *S. entellus* and *T. obscurus* and the apparent hybridisation in the wild of *S. entellus* with *T. johnii*.

The taxonomic relationship between these genera and another genus of middle-sized langurs, *Presbytis* Eschscholtz, 1821 of Indonesio-malayan origin, has also been the subject of dispute over time. A number of authors (Napier & Napier, 1967; Dao Van Tien, 1970; Lekagul & McNeely, 1988, to name a few) used *Presbytis* as the senior synonym for *Trachypithecus* and *Semnopithecus*. However, this genus is restricted by Weitzel & Groves (1985), and Groves (2001) based on cranial and dental evidence, locomotor morphology and behaviour (different brachial index), foraging adaptation and neonatal coat coloration.

In addition, there are enough morphological, behavioural, and molecular genetic differences to consider *Trachypithecus* as a full genus (Hill, 1934; Pocock, 1934, 1939; Weitzel & Groves, 1985; Eudey, 1987; Weitzel *et al.*, 1988; Groves, 1989, 2001; Roos *et al.*, 2001).

Groves (2001) placed the species of the genus *Trachypithecus* in five groups: *cristatus* group, *obscurus* group, *francoisi* group, *vetulus* group and *pileatus* group, whereas recent genetic studies show that the last two groups are more closely related to the genus *Semnopithecus* (see 2.3).

Species of three of these groups (*cristatus*, *obscurus*, *francoisi*) occur in Vietnam.

3.0.2   Morphology, ecology and behaviour

The predominant colouration of the species belonging to *Trachypithecus* in Vietnam is grey, dark brown or black. Most species have various white or yellowish markings on the head ("moustache", crest) shoulders or limbs.

All Vietnamese species have a pointed crest on top of the head distinct in the *francoisi*-group. The frontal whorl is poorly or not developed, and brow fringe is developed. The canine/sectorial complex is strongly sexually dimorphic (Groves, 2001).

The infant pelage colouration of *Trachypithecus* in Vietnam is a flamboyant orange (in some south Asian forms infants are grey or brown). Bright infant colouration is a common colobine feature.

Trachypithecus is mainly folivorous. Leaves are reported to comprise about 60% (Stanford, 1988 for *phayrei*; Bennett & Davies, 1994 for *Trachypithecus*) to 80% (Brotoisworo & Dirgayusa, 1991 for *cristatus*) of the diet. The remainder of the diet are shoots, fruits, flowers, and bark.

It appears that the species belonging to the *francoisi*-group share a preference for forest on limestone (karst) hills. This adaptation has been documented several times (e.g. Osgood, 1932; Tan, 1985; Dao Van Tien, 1989; Weitzel, 1992; Ratajszczak *et al.*, 1990, 1992; Weitzel & Thanh, 1992; Nadler, 1996b; Nadler & Ha Thang Long, 2000).

This association with karst also may explain why these species of the *francoisi*-group are allopatic. Weitzel (1992b) takes this idea further to hypothesize that the distinctive pelage among the forms of the *francoisi*-group, may be the result of isolated evolution in karst patches of Vietnam and China.

The association of *francoisi*-group langurs with karst is assumed to be due to three reasons (Nadler & Ha Thang Long 2000):

*   The vegetation and plant communities in limestone areas are species-specific food resources.

*   Caves used for sleeping provide good protection from predators.

*   Such caves offer lower temperatures in summer, higher temperatures in winter, and protection from rain and wind for species not well adapted to climatic conditions and change.

Results of the studies on the feeding plants of langurs show that while Delacour's and Cat Ba langurs prefer some tree species associated with limestone, they actually feed on a much broader spectrum of tree species (Nadler & Ha Thang Long, 2000). Langurs seem to have little to no difficulty using other vegetation types and plant communities in the same geographical region.

Langurs are mainly hunted inside the caves where they sleep, so their use of caves seems to pose a serious disadvantage to their survival. High hunting pressure at cave sites would seem to decrease the langurs' use of caves. On Cat Ba Island there is furthermore a lack of natural predators, most probably also in historical time, which makes it unlikely that caves are used by langurs for safety reasons.

Among the other limestone-associated langurs of the *francoisi*-group, the three northernmost species (White-headed langur, Cat Ba langur, and Francois' langur) are most closely connected with limestone. They regularly sleep in caves throughout the year. Temperatures in these species' range can drop below 10°C in winter (and even get close to freezing point). The Delacour's langur, which occurs further south, uses caves less frequently. In the Delacour's range, the cold period is shorter, and the average temperature is higher.

The southernmost two sub-species of the *francoisi*-group (*T. laotum laotum* and *T. l. hatinhenis*) occur in north-central Vietnam and central Lao PDR. According to new investigations (Duckworth *et al.*, 1999), these species occur in some areas without limestone cliffs, which also likely means without caves.

Burton *et al.* (1995) reported a close relationship between behaviour and temperature for *T. p. leucocephalus*. At temperatures lower than 10°C, the monkeys move swiftly, eat hastily, and then move to dry areas such as the underbrush in bamboo stands. At higher temperatures (11-30°C), they gather on ledges to sun, feed, and rest. To imitate natural conditions, the EPRC provides (isolated) sleeping boxes in the enclosures of its *Trachypithecus* species. EPRC observations confirm the close relationship between temperature and the use of sleeping boxes. With temperatures below 15°C the langurs use the boxes overnight and when temperatures are below 8-10°C, the langurs also stay inside the box during the day and exit only for feeding. Similar observations have been made at the Francois' langur breeding center in Wuzhou, China. This institution provides indoor enclosures for the langurs, and a heating system is used if the temperature drops down below 8°C (Mei Qu Nian *et al.*, 1998).

It seems that the primary reason langurs prefer limestone areas is due to the presence of caves and cracks that act as shelter against climatic conditions, especially temperature. Limestone-area caves with special micro-climatic conditions have been the main influence on the northern distribution of this langur group and also explain the limitation of the langurs' occurrence.

Langurs occasionally dash inside caves when feeling threatened. Lowe (1947) mentioned: "When in danger, it takes refuge in large holes in the limestone hills on which it lives." But this flight behaviour is probably a secondarily acquired behaviour.

The basic social organisation for this genus is the one male multi-female group. Other males form all-male bands. A male from such a band will eventually invade a one male multi-female group and replace the leader.

There is little information about the natural population density. The density is also dependent from the biomass production of the forest type and the size of the forest patches (Davies & Oates, 1994). Population densities in undisturbed habitats have been reported of 15 individuals/km² to 23-61/km² for *cristatus* (MacKinnon, 1986; Supriatna *et al.*, 1986), 57 individuals/km² for *obscurus* (Marsh & Wilson, 1981), 17-20/km² for *leucocephalus* (Chengming Huang *et al.*, 1998; Zhaoyuan Li *et al.* in print), 2,1-6,8/km² for *phayrei* (Fooden, 1971), and 8,2-64,3/km² for *geei* (Srivastava *et al.*, 2001).

Scarce information is available about the size of the home range for Vietnamese or closely related species. Zhaoyuan Li (pers. comm.) mentions 16-47ha (average of 30,5ha) for *leucocephalus*, Chivers (1973) 5-12ha for *obscurus*.

Studies on *Semnopithecus entellus* show a clear relationship between population density and home range size.

The troop size is species-specific, is related to population density, and ranges from 5 to 15 individuals.

Francois' langur males reach reproductive maturity at about 5 years of age, while females at the age of 4 years (Mei Qu Nian, 1998). The same goes for Hatinh and Delacour's langurs (Nadler, unpubl.).

For the northern Vietnamese species a reproductive peak was observed between January and June. The gestation period lasts 170-200 days (Davies & Oates, 1994; Mei Qu Nian, 1998; Nadler, unpubl.). The interbirth interval is 16-25 months (Davies & Oates, 1994), 19.1±2.2 months are mentioned for the Francois' langur (Mei Qu Nian, 1998), and 24.8+1.5 months for the Hatinh langur (Nadler, unpubl.). An interbirth interval of 8-9.5 months can occur if an infant does not survive (Francois' langur; Gibson & Chu, 1992)

The oestrus cycle is 24±4 days (Davies & Oates, 1994; Mei Qu Nian, 1998; Nadler, unpubl.)

Litter size is usually one, but twins are born occasionally (Hayssen *et al.*, 1993).

Within a one-male unit, infants are often exchanged between females. This behaviour is widespread among colobines and has inspired several hypotheses (Dolhinow & DeMay, 1982; Newton & Dunbar, 1994). These include the following: enabling the mother to forage, improving the maternal ability of the females of the troop; quickly integrating the infant into the group, and decreasing the resource-competition of the mothers for their own offspring. Newton & Dunbar (1994) noted, however, that the adaptive reason of this behaviour was not clearly demonstrated by field data.

The vocalization of the *francoisi*-group species varies somewhat from one another, and is quite distinctive from the "grey" species, Grey langur and Indochinese Silvered langur (Nadler, unpubl.).

3.0.3   Distribution

Species attributed to *Trachypithecus* exist in a disjunct range in Southern India and Sri Lanka and over a wide range from India through Myanmar and Southern China to Vietnam and Malaysia as well as the Indonesian Archipelago to Lombok.

Fooden (1996) argues that if congeneric taxa have complementary ranges, they can be treated as zoogeographic units. Following this, he divided *Trachypithecus* into two zoogeographic units in Vietnam.

The *francoisi*-group is restricted to karst formations in the north and central-north of Vietnam. *T. francoisi, poliocephalus, delacouri* and *laotum* change from north to south without their distributions overlapping. No intergradation has been documented. Information about sympatric occurrence of *hatinhensis* and *delacouri* (Brandon-Jones, 1995) was based on a misunderstanding and corrected (Brandon-Jones, 1996a).

T. crepusculus and *T. germaini* are distributed from north to south Vietnam, with *germaini* replacing *crepusculus* in the south, although the two species are separated by a large area.

3.1 Francois' langur

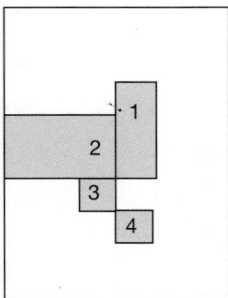

1. Francois' langur group. *T. Nadler*

2. Francois' langurs. *N. Rowe*

3. Francois' langur; ad. female. *T. Nadler*

4. Francois' langur; infant, five weeks old. *T. Nadler*

3.1 Francois' langur
Trachypithecus francoisi (Pousargues, 1898)

3.1.1 Taxonomy

Trachypithecus francoisi was first described by Pousargues (1898) from a specimen collected in southern Guangxi Province, China (22°24'N / 106°59'E). Six other taxa, each with a different pattern of white, grey or yellowish on a generally black or dark brown coat, have in the past usually been referred to as subspecies and formed the *francoisi* group. They include the Cat Ba langur, Delacour's langur, White-headed langur, Hatinh langur, Lao langur and Black langur.

Brandon-Jones (1984) argued that some of these taxa were distinctive at the species level.

Studies in the field and in captivity (Nadler, 1994, 1995c, 1997b), genetic research (see 2.3, Roos, 2000; Roos *et al.*, 2001) and parasitological findings (Mey, 1994) strongly suggest that some taxa, in the past usually defined as *francoisi* are distinct species. Groves (2001) listed all taxa as distinct species. Hence, this section refers to *francoisi* as a monotypic taxon.

3.1.2 Description

The pelage is glossy-blackish and only a narrow tract of slightly elongated white hair runs from the corner of mouth along the side of the face to the upper edge of ear pinna. There is a tall pointed crest on the crown of the head. The hair of the lateral parts of the body is long. Females have a depigmented pubic patch clothed with white to yellowish hairs.

Information about an all-black form without white cheeks is known from several areas where *francoisi* occur (Ratajszczak, 1990; Ratajszczak cit. in Brandon-Jones, 1995; B. Martin, pers. comm.; T. Nadler, pers. comm.). It is very unlikely that a second taxon exists widespread sympatrically with *francoisi*. Probably there are occasionally melanistic individuals like the melanistic morph "ebenus" in *T. laotum hatinhensis*. Observations in an area allegedly with the all-black form (s. locality Binh An Village, Chiem Hoa District, Tuyen Quang Province) and the fact that no specimen of the all-black form exists support the assumption that the all-black form is merely an incomplete observation. Langur sightings are mostly made close to the sleeping sites at dusk or dawn, and recognizing the white cheeks is often hardly possible.

External measurements (adult)

	n	mm	average	Source
Head/Body length				
male	4	510-635	565	Brandon-Jones, 1995
female	3	590-670	623	Brandon-Jones, 1995
Tail length				
male	4	805-900	863	Brandon-Jones, 1995
female	3	830-900	877	Brandon-Jones, 1995
Weight		kg		
male	11	6,4-7,85	7,05	Mei Qu Nian, 1998
female	1	7,2		Brandon-Jones, 1995
	8	5,5-7,9	6,69	Mei Qu Nian, 1998
?	2	6,5/7,2		Dao Van Tien, 1985

3.1.3   Distribution

The range of *T. francoisi* is limited to southern China and northern Vietnam. It is the most widespread species of the *francoisi* group. The northernmost part of its range is about 28°N latitude in Guizhou Province, China, and its southernmost extent is about 21°30'N latitude in Lang Son Province, Vietnam.

Distribution in China

Francois' langur is widely distributed throughout southern China but its range is highly fragmented (Zhang *et al.*, 1992). It occurs in more than twenty counties in western Guangxi Province and in four counties in southwestern Guizhou Province. Isolated fragments are found in three counties of northeast Guizhou Province. According to recent surveys the population in China totals 3,200-3,500 individuals. The species occurs in 16 reserves with a total of 1,900-2,000 individuals (Wang Yingxiang *et al.*, 1998).

Distribution in Vietnam

The historical range of *T. francoisi* in Vietnam may include the provinces of Ha Giang, Cao Bang, Lang Son, Bac Kan, Thai Nguyen, Tuyen Quang and the eastern parts of Lao Cai and Yen Bai. However, it seems to have been extirpated from most of its original distribution and can now only be found in isolated forest fragments.

3.1.4   Francois' langur records in Vietnam

Du Gia Nature Reserve (HA GIANG)
Special use forest: Nature reserve
Francois' langur status: Provisional occurrence, last report in 2001 (interview, hair sample), (Le Khac Quyet, 2001)

The provisional occurrence of this species in this locality is based on interview data collected in December 2000 to January 2001 during a FFI survey. A hair sample was also found in Lung Dam Village, Du Gia Commune, Yen Minh District (within the Nature reserve). The hair came from an animal shot in September 1999 (Le Khac Quyet, 2001).

Although this recent data represents a new distribution record for the species, the record is within the species range stipulated by Fooden (1996) and the limestone forest of Du Gia Nature Reserve provides suitable habitat for the species.

Local informants suggest 4-6 groups, each consisting of 3-7 individuals occur in the nature reserve. Further surveys are required to confirm the occurrence of the species in this area.

Cham Chu proposed Nature Reserve (TUYEN QUANG)
Special use forest: Proposed nature reserve
Francois' langur status: Provisional occurrence, last report in 2001 (interview)(Long & Le Khac Quyet, 2001)

Francois' langurs are reported without reference in Chiem Hoa District, Tuyen Quang Province (Ministry of Science, Techonology and Environment, 1992). Two surveys conducted in the area (Ratajszczak *et al.*, 1992; Dang Ngoc Can & Nguyen Truong Son, 1999) did not report this occurrence. A live animal was found in Nam Luong Village, Ham Yen District, however. It was bought from a villager in Khuon

DISTRIBUTION OF FRANCOIS' LANGUR (*Trachypithecus francoisi*) IN VIETNAM
BẢN ĐỒ PHÂN BỐ LOÀI VOỌC ĐEN MÁ TRẮNG Ở VIỆT NAM

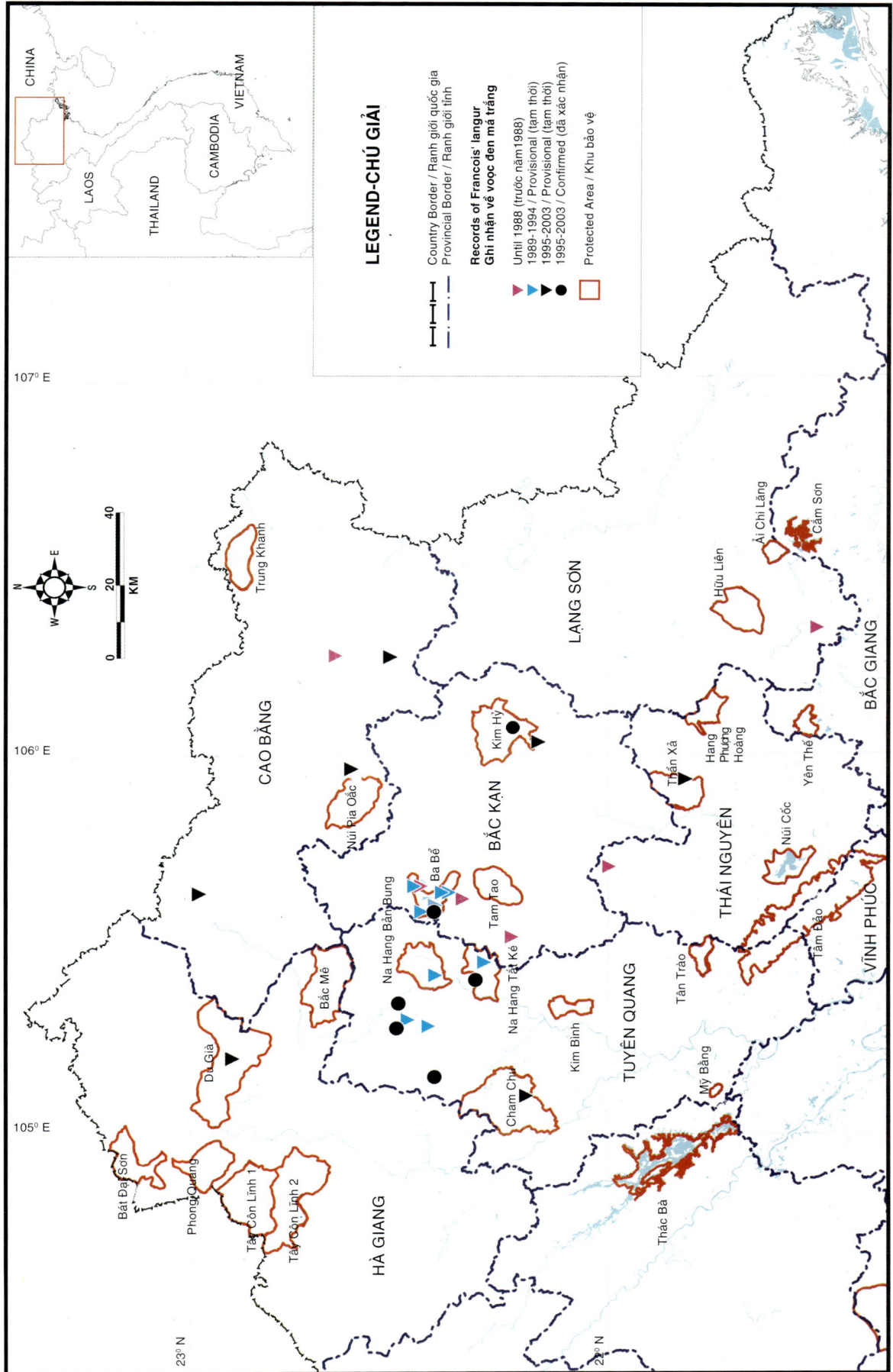

CHINA

VIETNAM

LAOS

THAILAND

CAMBODIA

LEGEND-CHÚ GIẢI

Country Border / Ranh giới quốc gia
Provincial Border / Ranh giới tỉnh

Records of Francois' langur
Ghi nhận về vooc đen má trắng

▶ Until 1988 (trước năm1988)
▶ 1989-1994 / Provisional (tạm thời)
▶ 1995-2003 / Provisional (tạm thời)
● 1995-2003 / Confirmed (đã xác nhận)

☐ Protected Area / Khu bảo vệ

107° E

106° E

105° E

23° N

22° N

CAO BẰNG

LẠNG SƠN

BẮC GIANG

BẮC KẠN

THÁI NGUYÊN

VĨNH PHÚC

TUYÊN QUANG

HÀ GIANG

Trung Khánh

Núi Pia Oắc

Na Hang Bản Bung

Ba Bể

Tam Tao

Na Hang Tát Kẻ

Bắc Mê

Du Già

Bát Đại Sơn

Phong Quang

Tây Côn Lĩnh 1

Tây Côn Lĩnh 2

Cham Chu

Kim Bình

Tân Trào

Mỹ Bằng

Thác Bà

Núi Cốc

Tam Đảo

Yên Thế

Hang Phượng Hoàng

Thần Sa

Kim Hỷ

Hữu Liên

Ải Chi Lăng

Cấm Sơn

40

20

0

KM

N

E

W

S

35

Pong 1 Village (Le Xuan Canh *et al.*, 2000). The location of capture of this animal was not given so the record remains provisional, although it is likely that it came from the adjacent limestone area of Ham Yen District. The animal died and was taken to the IEBR museum. Following interviews Long & Le Khac Quyet (2001) noted approximately three to four groups each numbering five to six animals in the area around Luong Pang. One brief sighting close to Dan Khao village, Phu Luu commune, likely represents this species but the sighting was not clear.

T. francoisi was described as common five years ago in Dan Khao, but severe hunting pressure has led its extirpation in some forest areas and population reduction in others (Long & Le Khac Quyet, 2001).

Nguyen Binh, Thach An and Bao Lam Districts (CAO BANG)
Special use forest: None
Francois' langur status: Provisional occurrence, last report in 2001 (interview), (Trinh Dinh Hoang, 2001; Le Khac Quyet & La Quang Trung, 2001)

Based on interview data a FFI primate survey of Cao Bang Province conducted in February/March 2001, indicated there may be some populations of the species in these districts. Further reports from Nguyen Binh District were obtained later in 2001 (Trinh Dinh Hoang, 2001). Most local informants in all three districts described a "Vuon duoi dai" or monkey with long tail, or refered to the "Khi khan quang" (white scarf around the neck). Furthermore, one skin was found in a household in Phieng Phat Village, Quang Lam Commune, Bao Lam District. Apparently the owner had shot the animal in 1999 near Du Gia Nature Reserve (Ha Giang) and reported that there are around 7-10 remaining individuals (Le Khac Quyet & La Quang Trung, 2001).

This new information confirmed the importance of the Du Gia NR and, in particular, the forest between the NR and Bao Lam District, Cao Bang Province. This forest may support a population of 10-20 individuals (Le Khac Quyet & La Quang Trung, 2001).

Interviews in Ha Lang District suggested that the species has been extripated from this area (La Quang Trung, 2001).

Ba Be National Park and proposed extension (BAC KAN)
Special use forest: National park
Francois' langur status: Occurrence confirmed, last evidence in 1999 (Nong The Dien, pers. comm. 2000)

In January 1927, Francois' langurs were observed by J. Delacour and W. P. Lowe (Lowe, 1947) in the vicinity of Ba Be lake. One female adult skull (ZMVNU 112) was collected in August 1967 by Truong Van La. One juvenile skull (ZMVNU 111, sex unknown) was collected by Le Ba Thai in August 1967 and was reported from Po Lu, Nam Mau Commune. Three skulls (male and two females, ZMVNU 128/3.120.26Ps, 87/3.35.11 and 113) were collected in March 1961 and February 1967 (collector unknown) in Ban Vai, Khang Ninh Commune. Two adult male skulls (ZMVNU 109 and 542) and one adult female skull (ZMVNU 110) were collected in July 1967 by Mr. Ha in the same locality. One juvenile skull (ZMVNU 117) was collected in March 1971 by Nguyen Van Chau in Nam Cuong Commune. One male skin (ZMVNU unnumbered) was collected by Vi Nguyen Thuyet in April 1983 in Ban Vai village (Fooden, 1996).

Information about the Francois' langur was obtained by Ratajszczak *et al.* (1990) in 1989. The team found the skin and skull of an animal shot in August 1989 near Dan Dang village (northwest of the park). Informants from this locality reported that a small group lived near their village, but was rarely seen. Other groups were also reported on the western shore of the central channel of the lake and in Dong Puong (=Puong Grotto) (northeast of the park).

From July to September 1994, a survey was carried out by Frontier (Kemp *et al.*, 1994). Local hunters and park staff reported the occurrence of Francois' langurs in Puong Grotto. A second Frontier survey took place from October to December 1996 (Hill *et al*, 1996a). One group of 4 to 5 animals was seen by the team near Nam Giai Village (west of the park) on two occasions in primary and disturbed forest at 750m a.s.l. The sighting of this species so close to the H'mong village of Nam Giai was seen by the authors as an encouraging sign.

In April 1999, Nong The Dien (pers. comm. 2000), Vice-director of the National Park, saw 6 animals on the west side of the southern lake. According to Nong The Dien (pers.comm. 2000) Francois' langur groups are still living behind Pac Ngoi Village, on the West side of the southern lake, along the Nang River, close to Dau Dang village and in Dong Puong area.

During the primate survey conducted by FFI in January 2000 in Bac Kan Province (Phung Van Khoa & Lormée, 2000), Francois' langurs were reported by local informants of Bang Phuc Commune, Cho Don District. Three interviewees claimed that they occurred on the boundary of their commune and Hoang Tri Commune (southwest of the proposed extension of the park)

North of Ra Ban Commune, Cho Don District (BAC KAN)
Special use forest: None
Francois' langur status: Provisional occurence, last report in 1989 (interview), (Ratajszczak *et al.*,1990)

The locality is reported by Ratajszczak *et al.* (1990) based on interviews during a primate survey conducted in north Vietnam in 1989. It was said to be very uncommon but could be still encountered on cliffs in the hills north of the commune, where the vegetation has not been exploited for fuel wood. The commune was not visited during the FFI survey in Cho Don District in January 2000 (Phung Van Khoa & Lormée, 2000). Three local villagers from Ra Ban, however, were interviewed but no data could be provided to support the current occurrence of *T. francoisi* in Ra Ban Commune.

Kim Hy Nature Reserve (BAC KAN)
Special use forest: Nature reserve
Francois' langur status: Occurrence confirmed, last evidence in 2000 (Ngo Van Tri & Lormée, 2000)

In March 1998, Geissmann & Vu Ngoc Thanh (2000) obtained several reports of Francois' langurs in the Kim Hy forest from locals. Based on interviews they reported that a group of 17 individuals was caught by hunters in their sleeping cave on the slopes surrounding camp 2 (22°11'N / 106°00'E) in 1997. Informants suggested that some animals were still living in the area because two animals had been caught since. This occurrence was confirmed when vocalisations were heard in the area. According to Geissmann & Vu Ngoc Thanh (2000) Francois' langurs are rare in this area. In November 1999, BirdLife and FIPI conducted a brief survey in Kim Hy Commune (Tordoff *et al.*, 2000b). Local informants (in three interviews) reported that langurs were fairly common. The last first-hand sighting was August-September 1998.

FFI carried out an initial short field survey of 10 days in January 2000 followed by three weeks of field work in this area during March 2000. Every village surrounding the forest was visited and a representative number of interviews with hunters were conducted in each. However, due to bad weather conditions, only a total of eight days was spent on surveys inside the limestone forest.

Limited time was spent in Kim Hy Commune, because, according to informants, wildlife has been pushed south as a result of the disturbance caused by gold mining activities and intensive hunting. All the local informants recognized the Francois' langur (twelve hunters were interviewed). A former sleeping site on a cliff close the path linking Kim Hy Village to the gold-mining village of Nam Djai was shown by villagers. According to the results of interviews, the Francois' langur no longer occurs in the vicinity of Kim Hy Village.

At least 20 interviews with local hunters were carried out in An Tinh Commune and Coc Xa-Coc Keng Villages (Con Minh Commune) and over 6 days of field surveillance in this forest section were undertaken (Ngo Van Tri & Lormée, 2000). Although considered rare, Francois' langur had previously been seen, heard and hunted. The field survey was able to confirm the presence of this species from fresh faeces found near a sleeping cave (22°11'09"N / 106°01'42.3"E). The cave was at about 25 meters elevation on a limestone cliff. Additionally, a skull was collected from the foot of the cliff. According to a Dao guide, this was the remains of an animal shot in 1998 or 1999 by a gold-miner. Another specimen, a left hand, was observed by the team in a hunter house near Kim Hy FPD station. The animal was reported to have been shot for meat in December 1999 in the forest between Con Minh and Kim Hy Communes.

A further survey in 2001 by FFI reported that a group of 10-15 langurs still survives in the area (La Quang Trung & Trinh Dinh Hoang, 2001).

In Cao Son Commune, logging and hunting appeared much more important than elsewhere in the area. Although most people could describe langurs, they all believed that they no longer occur in the commune forests.

Signs of hunting were observed everywhere and several informants revealed a local hunter at An Tinh Commune had shot 8 langurs in a sleeping site at Tao valley (not located) in 1999. At least one wildlife trader was visited by the team in each commune surrounding the forest bank. No captive or dead Francois' langurs were found, but informants reported that two individuals were sold by a trader from Lam Son Commune at the same time as the FFI survey in March 2000. According to the interviews, most of the animals are sent to China.

One informant in Kim Hy reported the occurrence of a second, much rarer form of langur with total black colouration (Ngo Van Tri & Lormée, 2000). The same information was reported by Tordoff *et al.* (2000b). However, other hunters did not mention this animal.

Na Hang Nature Reserve (TUYEN QUANG)
Special use forest: Nature reserve
Francois' langur status: Occurrence confirmed, last evidence in 1999 (B. Martin pers. comm., 2001)

In March 1992, Ratajszczak *et al.* (1992) reported that a small number (possibly up to ten groups) of Francois' langurs might survive on the slopes of Pac Ta Mountain, Tat Ke area. In Tat Ke area vocalizations were recorded by Boonratana & Le Xuan Canh (1994) during a survey conducted between September 1993 and February 1994.

One sighting of two animals was reported in August 1999 by Mr. Tinh (patrolling group) north of Pac Ta mountain, in Tat Ke sector (B. Martin, pers. comm. 2000).

The occurrence of the species in Ban Bung remains provisional and there is now little chance that langurs still occur in the sector.

A report of an enigmatic langur was provided by hunters from different villages interviewed separately in the Ban Bung area. This was a totally black, long-tailed monkey, with only a whitish V-mark on the chest. Local people who used Tat Ke and Ban Bung areas to hunt, could clearly differentiate this form from *T. francoisi.*

Lung Nhoi, West Bank of Gam River, Na Hang District (TUYEN QUANG)
Special use forest: None (proposed Na Hang extension; Momberg & Fredrickson, 2003)
Francois' langur status: Occurrence confirmed, last evidence in 2003 (Le Khac Quyet & Trinh Dinh Hoang, 2003b)

Reports of Francois' langurs living in very small numbers in the limestone area west of the Gam River were provided by interviews conducted by Ratajszczak *et al.* (1992) in March 1992. In addition, the team inspected a fresh skull in Ban Bo village and a living specimen caught one day before in Na Tong market.

A FFI/PARC survey confirmed the presence of Francois' langurs in Lung Nhoi, a small area (2000 hectare) of limestone pinnacles with highly fragmented forests between Thuong Lam, Xuan Tien, Xuan Tan, and Khuon Ha communes. Interviews suggest the presence of 5-7 groups, each comprised of 4-10 individuals. In March 2003 a group of 6-7 individuals in Keo Ac was confimed by sighting and vocalization. A second group was confirmed by vocalization in Phe Luong (Le Khac Quyet & Trinh Dinh Hoang, 2003a).

Duc Xuan area, East Bank of Gam River, Na Hang District (TUYEN QUANG)
Special use forest: None (proposed Na Hang extension; Momberg & Fredrickson, 2003)
Francois' langur status: Occurrence confirmed, last evidence in 2003 (Le Khac Quyet & Trinh Dinh Hoang, 2003b)

The presence of Francois' langurs has been confirmed in the limestone cliffs facing the Gam river in the Banh Trai area (Na Dua, Ban Phat, Da Den villages; Xuan Tien commune). Informants reported up to 10 groups with 5-10 individuals each. Trinh Dinh Hoang confirmed two groups by vocalization. Also in March 2003 Le Khac Quyet encountered two groups (possibly the same groups). One group of five individuals was sighted, a second group was confirmed through vocalization (Le Khac Quyet & Trinh Dinh Hoang, 2003b).

Than Xa forest, Vo Nhai District (THAI NGUYEN)
Special use forest: Partly included in Phuong Hoang-Than Xa Cultural and Historical Site
Francois' langur status: Provisional occurrence, last report in 1998 (interview), (Geissmann & Vu Ngoc Thanh, 2000)

Interviews conducted by Geissmann & Vu Ngoc Thanh (2000) in March 1998 reported the occurrence of Francois' langurs, with the most recent sighting made in February 1998. However another interview reported that monkeys no longer inhabit this area. It appears that the langurs, if they still occur in Than Xa, are now very rare.

Binh An Village, Chiem Hoa District (TUYEN QUANG)
Special use forest: None
Francois' langur status: Occurrence confirmed, last evidence in 2001 (Nadler, pers. comm., 2001)

One group of 9 animals was reported in June 1999 close to Binh An, but 6 animals were hunted during autumn 2000. Correspondingly three animals were observed in November 2000 and this was believed to be the total remnant population, although hunters report a group of approximately 30-40 individuals in the vicinity of Binh An.

The last evidence was a direct observation on 25-26/2/2001 of a group of 5-6 individuals in a sleeping place (Nadler, pers. comm., 2001).

In this area information indicates two different forms of langurs: one with white cheeks and a totally black one. Surprisingly, a totally black one has never been hunted.

The animals observed on the 25. and 26.2.01 were considered by hunters to be the all black form and the hunter accompanying the survey team insisted that the animals were all black. However, the observations were made at dusk and those member of the survey team equipped with binoculars could clearly spot the white cheeks.

Recent unpublished reports

During recent FFI surveys in Ha Giang Province, Francois' langurs have been reported from Xin Man, Bac Quang, Quan Ba and Bac Me Districts (Luong Van Hao, pers. comm.; Le Trong Dat pers. comm.). Further surveys are planned in these areas to assess the size of the population, the threats and to design threat mitigation strategies.

Locations where Francois' langurs are believed to be extinct:

Ban Thi Commune, Cho Don District (BAC KAN)

One juvenile female skull (IEBR 1474) was collected in November 1976 by Truong Minh Hoat, Ban Thi Commune. Two skins (male and female, ZMVNU 96 and 97) were collected in the same locality (date and collector unknown) (Fooden, 1996). FFI surveys conducted in October and November 1999 in the commune did not obtain any reliable information about the occurrence of Francois' langur (Dang Ngoc Can & Nguyen Truong Son, 1999).

One juvenile animal was collected in 1927 in Bach Thong district by Delacour and Lowe (Thomas, 1928)

Minh Tien, Huu Lung District (LANG SON)

Two adults (BMNH 1927.12.1.16 and 1927.12.1.17) were collected in January 1927 by J. Delacour and W. P. Lowe in Lang Son Province (Napier, 1985). One adult female skull (ZMVNU 95) was collected in June 1964 (unknown collector) in the Minh Tien locality (Fooden, 1996). Minh Tien Commune is located in an area south of Huu Lung District. The natural habitat around Minh Tien is largely fragmented and it is unlikely that it supports a significant population of langurs. According to the survey conducted by Frontier in Huu Lien Nature Reserve (north of Huu Lung District) *T. francoisi* may be extirpated in this area (T. Osborn, pers. comm., 2000).

Linh Thong Commune, Dinh Hoa District (THAI NGUYEN)

One juvenile male (ZMVNU 88/791) was collected in May 1967 by Pham Mong Giao, Linh Thong Commune (Fooden, 1996). The same individual is most probably also mentioned by Dao Van Tien (1990). No information was obtained in October 1999 during the FFI survey in the area (Dang Ngoc Can & Nguyen Truong Son, 1999).

Discussed records

Quang Ninh Province ca. 20°43'-21°40'N / 106°26'-109°04'E

The report of Francois' langurs in Quang Ninh Province (Ministry of Science, Technology and Environment, 1992) is considered as extremely unlikely by Fooden (1996). The species is also reported in the investment plan of Ky Thuong Nature Reserve, located in Quang Ninh Province. During a visit of the reserve by BirdLife and FIPI in November 1999, seven informants claimed that langurs did not occur in the area (Tordoff *et al.*, 2000b).

Phong Quang Nature Reserve (HA GIANG) ca. 22°50'-23°04'N / 104°50'E-105°01' E

Francois' langurs are reported in the investment plan for Phong Quang Nature Reserve, Ha Giang Province (Anon., 1997b). However, no reference is given.

Cao Bang vicinity (CAO BANG)

Billet (1896) reported about an animal from Cao Bang city but without clear information about its origin.

3.1.5   Status

The population of *T. francoisi* is decreasing throughout its range, due to poaching and habitat degradation. The populations are now considerably fragmented.

According to the preliminary survey conducted in October 1998, the distribution of Francois' langur seems to be dwindling in China (Li Zhaoyuan, pers. comm., 2000). For instance, in Encheng Reserve (Daxin County, Guangxi Province), which was the major reserve protecting the species in 1988 and where all the observations by Chinese authors were made, no animals could be seen. According to local hunters, all langurs were removed due to poaching.

Tan (1985) gave an estimation of 5,000 to 6,000 Francois' langurs in China; Wang Yingxiang *et al.* (1998) 3,200 to 3,500 individuals. Regarding the level of human pressure on the population and its natural habitat, this number may currently be significantly lower.

Northern Vietnam suffers more from forest destruction than any other part of the country. The causes of this destruction are primarily conversion to agricultural land, fuel wood collection and timber extraction. The remaining natural habitat is now restricted to scattered, fragmented areas. Although several forest blocks may have sufficient size to support a significant number of Francois' langurs, such as Na Hang or Kim Hy forests, the species seems to live in very low densities. Hunting is obviously the major threat to its survival in the near future.

Francois' langurs are hunted both for meat and to be used in traditional medicine. The proximity of the Chinese border, where a significant number of animals are sold to supply local restaurants and traditional medicine producers, increases the pressure exerted on the population. The FFI surveys in Kim Hy forest (Ngo Van Tri & Lormée, 2000), revealed that every locality surrounding the forest had at least one wildlife trader. The results of the interviews showed that most of the animals, including *T. francoisi*, were destined for the Chinese market.

Mining activities, common in the limestone areas where Francois' langur occurs, increase the human pressure on the species and its natural habitat. Gold is extracted in Kim Hy and Na Hang and two companies exploit zinc and aluminum ores in Ban Thi. The arrival of hundreds of miners has multiplied the number of inhabitants and consequently the number of forest product users in the region. In addition, the clearance of large areas where ores are found causes signficant damage to the natural forest. FFI surveys in Na Ri District confirmed the harmful effect of mining activities on the environment (Phung Van Khoa & Lormée, 2000; Ngo Van Tri & Lormée, 2000). Every informant interviewed in Kim Hy Commune suggested to the teams that wildlife density was seriously depleted around the gold miners villages in the northern part of Kim Hy forest.

From over ten years of surveys in northern Vietnam, very few sightings of Francois' langurs are recorded in the scientific literature. This is less than any other langur species. It is likely that the species has been almost extirpated from its historical range. Ratajszczak *et al.* (1990) surveyed a large part of the provinces included in its distribution area, but obtained little information. Tordoff *et al.* (2000b) did not record the species in two locations of Cao Bang Province, Trang Hen (Tra Linh District) and Trung Khanh Nature Reserve (Trung Khanh District) although there exists suitable habitat for the species. During the eight surveys carried out by FFI in Tuyen Quang, Bac Kan, Ha Giang and Cao Bang Provinces to assess the primate status, confirmed evidence of its occurrence was only collected in Kim Hy Nature Reserve, Ba Be National Park, Lung Nhoi (West bank of Gam river), and Duc Xuan (East bank of Gam river). Despite the number of interviews conducted throughout the surveyed area, provisional reports were obtained only in another six places and the species was noted as very rare and shy.

Based on the interview data presented, the population is estimated to be less than 300 individuals (97-294) in Vietnam, devided into at least 10 sub-populations.

A problem for the long term survival of the species is that the population is now very fragmented and no interbreeding between the different groups is possible. In the future this may result in a genetic degradation.

The global conservation status in the IUCN Red List of Threatened Species (Hilton-Taylor, 2000) is **"Vulnerable" VU** A1cd, A2cd, C2a. Although there are little new information from China, the species may be very close to **"Endangered"** status.

The species is currently considered as **"Endangered"** in Vietnam (Pham Nhat *et al.*, 1998), but **"Critically Endangered" CR** A1cd, C2a best reflects the reality of the situation in Vietnam, because it can be estimated that no subpopulation contains more than 50 mature individuals.

3.1.6   Recommendations for conservation in Vietnam

3.1.6.1   Conduct further field status surveys

No area in Vietnam is known to possess a significant population of Francois' langur. However, it is possible that other populations occur in more remote forests, which have not yet been surveyed. With regards to the current status of the species, urgent measures should be taken to protect the species in the areas in which the presence of the species can be confirmed.

Although few suitable habitats remain in the historical range of *T. francoisi*, urgent surveys should be conducted to locate larger populations, particularly in Ha Giang Province. Currently, only one provisional record of the species is available about the status of this species in the province.

The occurrence of Francois' langur was recently confirmed in Kim Hy forest by FFI in March 2000. However, no sightings were obtained and no reliable estimation could be provided by local hunters. Although it appears that the species is uncommon in Kim Hy, being one of the last localities in which it can be found, the area should be surveyed intensively, mainly in Con Minh and An Tinh forests.

Na Hang Nature Reserve is known to hold a population of Francois' langur. However, the area has never been surveyed with an emphasis on this species, although the species' occurrence is confirmed by an opportunistic sighting in Chiem Hoa district (Nadler, pers. comm., 2001). A survey should be conducted immediately on Pac Ta mountain and Cham Chu proposed Nature Reserve, where a population allegedly lives, in order to assess its status in the reserve (Long & Le Khac Quyet, 2001).

3.1.6.2   Expansion of protected area network

The investment plan to establish Kim Hy Nature Reserve has recently been approved by MARD, but no on-the-ground conservation management exists. Furthermore, the proposed boundaries are unable to provide any significant protection for wildlife in general and Francois' langurs in particular. Indeed about 27% of the natural forest is not included in the plan and 43% of the proposed protected area is comprised of settlements, fields and grasslands (Tordoff *et al.*, 2000b).

The Lung Nhoi area west of the Gam river and the Duc Xuan/ Sinh Lonh area north of Na Hang Nature Reserve with forest contiguous with the Tat Ke sector should be included within Na Hang Nature Reserve or designated as seperate Species and Habitat Conservation Area (Momberg & Fredrickson, 2003)

3.1.6.3   Reinforcement of conservation in protected area network

Ba Be National Park, Du Gia Nature Reserve, Na Hang Nature Reserve and Cham Chu Nature Reserve are the only protected areas known to possess a population of Francois' langur. The status of the species should be reassessed and special conservation measures (e.g. hunter targeted awareness programmes, creation of patrolling groups) should be undertaken. These measures should be applied in the proposed extension of Ba Be NP and Na Hang Nature Reserve where no on-the-ground conservation management exists yet.

3.1.6.4   Other options for conservation

Currently, no information is available to conclude with any degree of certainty that a viable population of *T. francoisi* remains in Vietnam. The last hope for the survival of the species could be a captive breeding programme for future reintroduction. The existing international captive management programme (D. Pate, pers. comm., 2000) for *T. francoisi* should be expanded. An extensive and successful breeding project for the species runs at the Wuzhou Breeding Center in China (Mei Qu Nian, 1998).

3.2. Hatinh langur

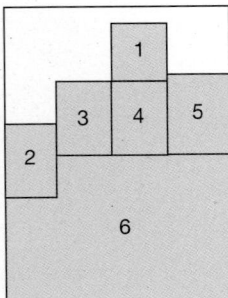

1. Hatinh langur; ad. male. *T. Nadler*

2. Hatinh langur; ad. male. *T. Nadler*

3. Hatinh langur; infant female, four weeks old. *T. Nadler*

4. Hatinh langur; infant male, six month old. *T. Nadler*

5. Hatinh langur; black morph "ebenus", ad. male. *T. Nadler*

6. Hatinh langurs. *J. Holden*

3.2   Hatinh langur
Trachypithecus laotum hatinhensis (Dao, 1970)

3.2.1   Taxonomy

The Hatinh langur was described by Dao Van Tien (1970) as a sub-species of *T. francoisi*, based on two specimens collected in 1942 and 1964. The type was reported from Xom Cuc, in Quang Binh Province near its border with Ha Tinh. (The occurrence of Hatinh Langur in the latter province has never been confirmed). Most authors have agreed with this taxonomy. Groves (2001) placed the langur as a distinct species. However, a recent taxonomic review, based on genetic evidence, would separate *hatinhensis* from *francoisi* and include it as subspecies of *T. laotum* (see 2.3, Roos, 2000; Roos *et al.*, 2001).

T. l. hatinhenisis is a dimorph taxon with (partly) geographically separated but not reproductively isolated forms and intergradation.

In 1924, F. R. Wulsin (Thomas, 1928) collected the skin and skull of an all-black langur, now in the Smithsonian Museum (USNM 240489). The locality "French Indo-China" does not indicate its true provenance. Brandon-Jones (1995) speculates that the animal originated from the Fan Si Pan area in North Vietnam and allocates it to the Indonesian Ebony langur *Semnopithecus auratus*, which is also black, as a new subspecies *S. a. ebenus*.

In January 1998 the EPRC received an all-black langur which came presumably from the Hin Namno area in Lao (Nadler, 1998).

Anatomical features, behaviour and vocalization of this animal show strong affinities to the Hatinh langur. The Black langur does not differ in colouration, hair structure, whorls and shape of the crest from the Hatinh langur. The moustache and the cheeks are black but the hair structure is different, with the hairs being more wavy and marking the parts between the corner of the mouth and the ears. Brandon-Jones' (1995) detailed description of the hair structure, hair direction and colouration of the specimen in the Smithsonian Museum corresponds closely to the animal at the EPRC and differs clearly from the Ebony langur.

The geographical separation also suggests that the Black langur is not a subspecies of the Ebony langur, which is distributed on the islands Java, Bali and Lombok. Finally the itinary of Wulsin's expedition suggests that the type was actually collected in the Hin Namno area as well, and not in northern Vietnam as originally assumed.

Based on the corresponding morphology and molecular genetic of the type specimen and the living animal at the EPRC, they are most probably of the same origin.

Recent genetic studies have shown that the Black langur is only an melanistic morph of the Hatinh langur (see 2.3). This is also supported by field observations, in that both morphs are present in the same region (Ruggieri & Timmins, 1995).

3.2.2   Description

The Hatinh langur differs from Francois' langur in the extent of the white cheek stripe behind the ear onto the nape. Other differences include the whorls on the head and the shape of the crest, and a white moustache connecting the white cheek stripes (Nadler, unpubl.). The nominate form *T. l. laotum* (Thomas, 1921) from Ban Na Xao, Lao PDR (ca. 17°30'N) has a white head with a black crest and brow band. Furthermore, an all-black form *T. laotum ebenus* (see below) has also been described from this region.

External measurements (adult)

	n	mm	average	Source
Head/Body length				
male	2	560-590	575	EPRC
	1	665	665	Brandon-Jones, 1995
"ebenus"	1	620	620	EPRC
female	3	540-570	556	EPRC
	1	500	500	Brandon-Jones, 1995
Tail length				
male	2	820-870	845	EPRC
	1	810	810	Brandon-Jones, 1995
"ebenus"	1	940	940	EPRC
female	3	780-900	817	EPRC
	1	870	870	Brandon-Jones, 1995
Weight		kg		
male	2	8,2-8,7	8,45	EPRC
	1	8,0	8,0	Brandon-Jones, 1995
"ebenus"	1	10,3	10,3	EPRC
female	4	6,4-8,0	7,2	EPRC

3.2.3   Distribution

Although the historical range of *hatinhensis* may once have been much more extensive (Nadler, 1996c), it seems currently restricted to the limestone areas in the west part of Quang Binh Province and to a lesser extent in the eastern part of Khammouan Province (Lao PDR).

The Phong Nha-Khe Bang area in combination with the Hin Namno area of Laos has probably the largest remaining single population of animals of this taxon globally (Timmins *et al.*, 1999).

The black morph was only reported from the limestone complexes Phong Nha-Ke Bang (Vietnam), Hin Namno NBCA and Khammouan Limestone NBCA (Lao PDR).

Distribution in Lao PDR

Sightings of Hatinh Langurs in Lao PDR are occasional, and their distribution in this country appears limited. This taxon has only been recorded for certain in the Phou Vang area, Nakai Nam Theun NBCA (Robichaud, 1999). In Hin Namno NBCA, some individuals showed features of the head pelage tending towards *T. l. hatinhensis*, but most appeared black-headed (Ruggieri & Timmins, 1995; Timmins & Khounboline, 1996; Duckworth, 1999; Walston & Vinton, 1999). This is apparently the predominant form in this area, although some individuals, showing features of the head pelage tending towards to *T. l. hatinhensis*, also occur here (Walston & Vinton, 1999; Duckworth *et al.*, 1999).

A black-headed form was also reported in the southern extremity of Khammouan limestone NBCA (M. F. Robinson *in litt.*, 1999; cited in Duckworth *et al.*, 1999) where *T. l. laotum* commonly occurs. During a 1998 survey in this area all identifiable sightings were of typical *laotum* (Duckworth *et al.*, 1999).

Distribution in Vietnam

Le Xuan Canh (1993) photographed a typical Hatinh langur in June 1992 in Phong Nha village. Until then, the Hatinh langur was only known from its type specimens. Since then, a number of records have increased our knowledge about the distribution of this taxon.

The only confirmed current occurrence are in the Phong Nha-Ke Bang National Park and Tuyen Hoa District (Quang Binh Province). There is no evidence of its occurrence in Ha Tinh Province and no data are available for Quang Tri Province.

The only sighting of an all-black langur which could be related to the animals observed in Hin Namno NBCA, was made during the survey carried out by FFI from July to September 1998 in Phong Nha-Ke Bang National Park (Nguyen Xuan Dang *et al.*, 1998). At least one individual of a group living in Thung Ba Dau (17°36'N / 106°17'E) was observed twice by the team in July 1998 at the same sleeping site. This form was reported in interviews with local people throughout the area

The record of all-black langurs in Phong Nha-Ke Bang area by an FFI team in 1998 was not previously documented. This suggests that earlier records should be reviewed to determine which form was identified. However, it is unlikely that pelage features were carefully checked during each sighting. Furthermore, it seems that data from some sightings are still unpublished (Timmins *et al.*, 1999).

3.2.4   Hatinh langur records in Vietnam

Phong Nha-Ke Bang National Park (QUANG BINH)
Special use forest: National park
Hatinh langur status: Occurrence confirmed, last evidence in 2003 (Nadler, pers. comm., 2003)

In June 1992, Le Xuan Canh photographed a single adult male collected alive by a local hunter in Phong Nha village, Bo Trach District (Le Xuan Canh, 1993). This was the first evidence of the existence of this subspecies since description of the species by Dao Van Tien (1970). In 1994, one juvenile female was taken to Xuan Mai Forestry College from Minh Hoa market (skin preserved in FCXM, unnumbered).

Between 1993 and 1998 the EPRC received 16 Hatinh langurs mostly confiscated from the animal trade in Quang Binh province and most probably caught in the Phong Nha-Khe Bang area (Nadler, 2000).

During June and July 1994, a survey for endemic pheasants was conducted by NWF, IUCN, WWF and BirdLife (Lambert *et al.*, 1994). One group of langurs was observed, sleeping every night on a limestone cliff. The animals were not clearly identified as Hatinh or Black langurs. Local people reported that langurs were widely distributed throughout the area.

In 1995 and 1996, three surveys were carried out in Thuong Hoa area (Minh Hoa District) and in Phong Nha area (Bo Trach District) by Pham Nhat *et al.* (1996a) with a special emphasis on the Hatinh langur. Five groups were observed in June and July 1995, four in November 1995 and three in June 1996. In addition, the species was reported, based on interviews, in Hoa Son, Minh Hoa District and an old sleeping site was checked by the team.

In March 1998, between 10 and 20 langurs were seen from road No. 561 to the south of Cha Lo Village at a sleeping site but not clearly identified as Hatinh langurs (Sterling in lit. cited by Timmins *et al.*, 1999)

Field surveys were conducted from July to October 1998 by FFI (Timmins *et al.*, 1999). Several groups of Hatinh langurs were observed, with the number of animals varying between 4 and 8 animals. In July 1998, one group was detected in a sleeping site in Hung Ba Dau (ca. 17°36'N / 106°17'E), though the

DISTRIBUTION OF HATINH LANGUR *(Trachypithecus laotum hatinhensis)* IN VIETNAM
BẢN ĐỒ PHÂN BỐ LOÀI VOỌC HATINH Ở VIỆT NAM

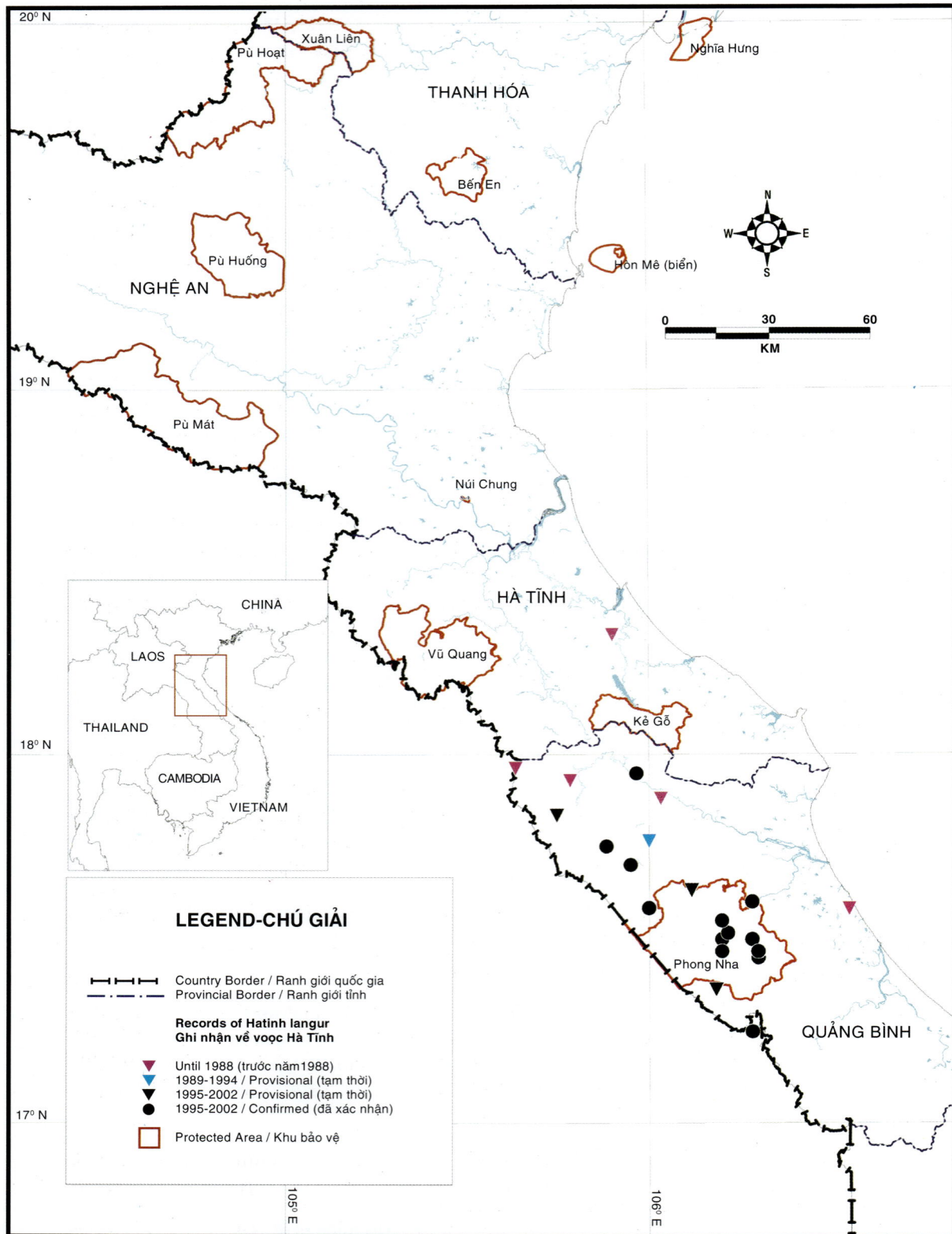

LEGEND-CHÚ GIẢI

Country Border / Ranh giới quốc gia
Provincial Border / Ranh giới tỉnh

Records of Hatinh langur
Ghi nhận về voọc Hà Tĩnh

▼ Until 1988 (trước năm 1988)
▼ 1989-1994 / Provisional (tạm thời)
▼ 1995-2002 / Provisional (tạm thời)
● 1995-2002 / Confirmed (đã xác nhận)

☐ Protected Area / Khu bảo vệ

primates were not seen. The same month, one group of at least 4 individuals was observed near a cliff close to basecamp 1 in the Suoi Chay valley (17°30.847'N / 106°12.776'E). At the same time and same location several loud calls were heard in an opposite direction. In July, one group of at least 5 animals was seen regularly in a sleeping cave at the Eo Cap site (17°32.638'N / 106°12.687'E). In August 1998, one group of at least five members was heard and briefly observed in a sleeping cave at the Hung Lao site (17°32.827'N / 106°12.809'E). White cheeks could be seen. Two animals with white cheeks were seen in October 1998, in the Ta Ty area of Cha Lo Village (ca. 17°43'N / 105°47'E). A single animal was observed in the south of Hang Ca Tuc area (17°35'N / 106°00'E) in October 1998. Head pelage was not seen. In September and October 1998, vocalizations were heard twice to the east of Cha Noi (17°38'N / 106°05'E), once to the south of Phu Nhieu (17°40'N / 106°02'E), seven times in the Suoi Chan Loong valley (from four different places) (17°37'N / 106°00'E), eight times in the Hang Ca Tuc area (from at least four locations) (17°35'N / 106°00'E), once from the southern side of Mo O valley (17°40'N / 105°57'E), four times in the Suoi Mo Sang valley (from different locations on one occasion) (17°38'N / 105°55'E), and three times in the Suoi Cat valley (from seemingly three different locations on one occasion) (17°43'N / 105°47'E).

According to nature reserve rangers, at least three groups live on the cliffs surrounding Suoi Chay-river where they are easily encountered in the early morning. Two animals were seen by FFI staff in this location during a short visit in June 2000 (Lormee, unpubl.). This river valley is currently being subjected to heavy disturbance as a road is being built along the side of the river. The effect of this development on the primates is currently unknown.

According to Timmins *et al.* (1999), very few sightings and vocalizations were recorded and this suggested a reduction of the population. The same authors hypothesise that arrival and departure from sleeping sites in the dark, as well as the probable reduction of the vocal activity (compared with areas of comparable habitat) may reflect a change of behaviour in response to the human pressure.

Khe Net Nature Reserve, Tuyen Hoa District (QUANG BINH)
Special use forest: Nature reserve
Hatinh langur status: Occurrence confirmed, last evidence in 2000 (Le Trong Trai, pers. comm., 2000)

Tuyen Hoa District is the origin of the type and paratype described by Dao Van Tien (1970). The type, one adult female (ZMVNU, 90) was collected in December 1942 by Chau in Xom Cuc locality. The paratype, one male skin, was collected in 1964 in Ninh Hoa Commune during a survey organised by the State Committee for Science and Technology (museum and collector unknown).

During their survey for Hatinh langur in 1995 and 1996, Pham Nhat *et al.* (1996a) could not find any information about this species in the area.

However, Le Trong Trai (pers. comm., 2000) observed three langurs in June 2000 in the limestone forest of Kim Lu area, Kim Hoa Commune (17°56,917'N / 105°57,937'E). The animals were too far away to actually see whether the head pelage was consistent with *hatinhensis* or with the black morph "ebenus". This could be the first record of the Hatinh langur in the area. Local hunters reported that 20 to 30 animals still inhabited the area.

Locations where Hatinh langurs are believed to be extinct

Ba Ren and Truong Son, Bo Trach District (QUANG BINH)

No information was collected by Pham Nhat *et al.* (1996a) in the area during a survey with a special emphasis on Hatinh langur in 1995 and 1996. The species is believed to be locally extinct (Pham Nhat *et al.*, 1996a).

Huong Son District (HA TINH)

No information was collected by Pham Nhat *et al.* (1996a) in the area during a survey with a special emphasis on Hatinh langur in 1995 and 1996. The species is believed to be locally extinct (Pham Nhat *et al.*, 1996a).

Discussed records

Nhu Xuan District (THANH HOA) ca. 19°39'N/105°29'E and Tan Ky District (NGHE AN) ca. 19°05' N/ 105°25'E

The information about the occurrence in these areas (Nadler, 1996e) is based on stuffed animals which were found in hunters' houses. The hunters said that they had killed the animals in their fields close to limestone outcrops which made an occurrence reliable. The hunters believed that the killing of protected animals on their own land constituted extenuating circumstances. Subsequently, it was discovered that the animals were actually hunted in Quang Binh Province, most probably in the Phong Nha area.

Vu Quang Nature Reserve, Huong Khe District (HA TINH) ca. 18°09'-18°25'N / 105°16'-105°36'E

Hatinh langur was reported in the draft management plan of Vu Quang Nature Reserve (McKinnon, 1992). The locality is also reported in the reserve by the VRTC based on interviews conducted in July-August 1997 (VRTC 1997). However, this reports has never been confirmed. Pham Nhat *et al.* (1996a) heard the same reports but local informants could not identify the pictures of this primate taxon (Pham Nhat *et al.*, 1996a). The latter authors concluded that this occurrence is questionable. Limestone hills, apparently preferred by the langurs, do not exist at the site.

Pu Huong Nature Reserve (NGHE AN) ca. 19°15'-19°29'N / 104°43'-105°00'E

Hatinh langurs were reported without reference in the investment plan of Pu Huong Nature Reserve (Anon., 1995b). Regarding the geographic situation of the reserve, such occurrence is questionable. Hatinh langurs were not recorded in 1995 during the field survey of Frontier (Kemp & Dilger, 1996).

Pu Mat Nature Reserve (NGHE AN) ca. 18°46'-19°08'N / 104°24'-104°59'E

Hatinh langurs were reported by Le Hien Hao (1973) in the Pu Mat area without references. However, the existence of this species in the reserve now seems doubtful. This former record was probably from forest on limestone karst in or near the buffer zone of Pu Mat. These areas are now surrounded by intensive agriculture and, due to hunting and trapping, unlikely to support any viable large mammal populations (Johns, 1999).

Le Thuy District (QUANH BINH) ca. 17°15'N/106°30'E

The occurrence is mentioned in the *Red Data Book* (Ministry of Science, Technology and Environment, 1992) without detailed information.

Bach Ma National Park (THUA THIEN-HUE) ca. 16°05'-16°16'N / 107°43'-107°53'E

Hatinh langur was reported in the management plan for Bach Ma National Park (Anon., 1990b). This occurrence could not be confirmed by Pham Nhat *et al.* (1996a). The latter authors noted that if the species was present at such a latitude, it was probably a relic population.

Kon Cha Rang Nature Reserve (GIA LAI) ca. 14°26'-14°35'N / 108°30'-108°39'E

Lippold & Vu Ngoc Thanh (1995a) reported the occurrence of *T. l. hatinhensis* in Kon Cha Rang Nature Reserve under difficult viewing conditions. Pham Nhat *et al.* (1996a) considered this occurrence as doubtful. They suggest that if the taxon was present in this latitude, it was probably a relic population. This taxon was not reported during a survey for large carnivores in the area (Le Xuan Canh, 1993). The presence of Hatinh langurs in the Tay Nguyen Plateau would be a major extension of the range of this taxon and should be investigated further. This record is not regarded as confirmed in this status review.

Krong Trai Nature Reserve (PHU YEN) ca. 13°01'-13°10'N / 108°46'-108°57'E

T. francoisi (probably in reference to *hatinhensis*) was reported based on interviews in Krong Trai Nature Reserve by the investment plan conducted by FIPI (Anon., 1990a). This record cannot be considered as reliable.

Cat Tien National Park (DONG NAI, BINH PHUOC and LAM DONG) ca. 11°21'-11°48'N / 107°10'-107°34'E

T. francoisi (probably in reference to *hatinhensis*) was reported in the Management Plan of Cat Tien National Park (Anon., 1993b). This record cannot be considered as reliable.

3.2.5 Status

The current range of *T. l. hatinhensis* is concentrated in Quang Binh Province. As with other primates in Vietnam, the main threat is hunting. Although a large area of suitable habitat remains in Phong Nha, the population density appears very low.

Shooting primates is quite common throughout the range of the Hatinh langur. Animals are killed for meat as well as for traditional medicine production and wildlife trading. In 1991, the Command Board of Quang Binh Province ordered the return of all rifles distributed to the local militia during wartime. However, this was not completely implemented and poaching still continues (Pham Nhat *et al.*, 1996a). Snaring is reported to be the predominant hunting method in Phong Nha-Ke Bang (Timmins *et al.*, 1999). These authors also suggest that snaring intensified since 1996, with the increasing demand for live animals. Hunting with rifles appears to have diminished since 1995, probably due to gun legislation controls, further confiscation of guns by local authorities and a decrease of hunting success.

Hunting seems primarily commercially oriented in Phong Nha-Ke Bang. Timmins *et al.* (1999) report primate hunting expeditions, where the dried carcass and meat are prepared within the forest. The use of bones to make valuable traditional medicine ("cao khi") is high. In addition, the trade in living animals is intensive, as shown by the number of individuals confiscated and now kept at the EPRC in Cuc Phuong National Park.

The natural habitat in Quang Binh province is now restricted to a band of forest near the Lao border. Forests are heavily degraded in Tuyen Hoa and northern Minh Hoa Districts (Pham Nhat *et al.*, 1996a). In Phong Nha-Ke Bang forest, habitat conversion is relatively low compared to other regions in Vietnam and is limited to the most accessible areas. It is predominantly the result of timber and firewood extraction. Little evidence was found for exploitation of resources other than timber and wildlife.

Since 2002, road construction through the middle of Phong Nha-Ke Bang National Park has increased for hunting and illegal logging. Dynamiting for the road, through the karst has let to severe disturbance of the Hatinh langur population along the Chay river.

Based on interviews and field observations, Pham Nhat *et al.* (1996a) estimated 520 to 670 Hatinh langurs in Phong Nha-Ke Bang, 50 to 70 of which were in Hoa Son area, 250 to 350 in Thuong Hoa area and 200 to 250 in Phong Nha area.

The Hatinh langur is listed in the *IUCN Red List of Threatened Species* (Hilton-Taylor, 2000) as a subspecies of *T. francoisi* and the global conservation status classified as **"Endangered" EN** A1cd.

This taxon is now identified as a subspecies of T. laotum. In Vietnam the Hatinh langur is considered as **"Endangered"** (Pham Nhat et al., 1998) and should be listed for Vietnam with the criteria **EN** A1cd, A2cd, B2bc, C1.

The Black langur is listed the *IUCN Red List of Threatened Species* (Hilton-Taylor, 2000) as a subspecies of *T. francoisi* and the global conservation status classified as **"Data Deficient"**. In Vietnam the Black langur is also considered as **"Data Deficient"** (Pham Nhat et al., 1998). The taxon is now identified as a melanistic morph of *T. laotum hatinhensis*.

3.2.6   Recommendations for conservation in Vietnam

Hatinh langurs were the subject of special surveys to assess their status in Vietnam in 1995 and 1996 (Pham Nhat *et al.*, 1996a). The compilation of these data with the results of the most recent surveys conducted in Vietnam allows us to propose the following measures.

3.2.6.1.   Conduct further field status surveys

The range of the species may extend to the south and north of Phong Nha-Ke Bang. Further surveys are needed to identify other populations. Particular areas of interest may be the limestone forest of southern Quang Binh Province and the forest of Giang Manh mountain north of Ke Bang.

3.2.6.2.   Reinforcement of conservation in protected area network

The conservation of the Hatinh langur is closely linked with the future of Phong Nha-Ke Bang Proposed National Park. The recent extension of the protected area covers a much larger part of the current range of the species (about 150,000 ha).

However, boundary demarcation, assignment and training of rangers, infra-structural development as well as the development of a National Park Management Plan should happen as soon as possible, since the hunting pressure is increasing. Areas with easy access due to the recent construction of the Ho Chi Minh Highway and the Lao road link through the park need to be targeted immediately for increased patrolling and enforcement. Four globally threatened primate species are confirmed for this reserve. Hence Phong Nha-Ke Bang should be considered as a global priority site for primate conservation.

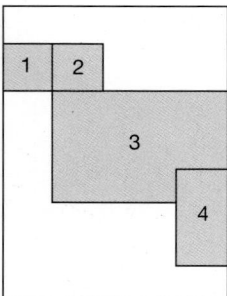

1. Cat Ba langur; ad. female. *T. Nadler*

2. Cat Ba langur; infant male, three weeks old. *T. Nadler*

3. Cat Ba langur group. *S. Kobold*

4. Cat Ba langur; infant female, about seven month old. *T. Nadler*

3.3   Cat Ba langur
Trachypithecus poliocephalus poliocephalus (Trouessart, 1911)

3.3.1   Taxonomy

The Cat Ba or Golden-headed langur was described as a new species, *poliocephalus*, of the genus *Semnopithecus* by Trouessart (1911) on the basis of a single adult female skin. The skin was donated to the Museum National d'Histoire Naturelle, Paris. Subsequently, this primate taxon has been placed as a species in several genera: *Pithecus poliocephalus* (Thomas, 1928), *Trachypithecus poliocephalus* (Pocock, 1935), and *Pygathrix poliocephalus* (Rode, 1938).

Ellerman and Morrison-Scott (1951) grouped this taxon with the Delacour's langur and the Lao langur as a subspecies of the Francois' langur: *Presbytis francoisi poliocephalus*. Many authors follow this decision although they sometimes alter the generic classification to either *Trachypithecus francoisi poliocephalus* (Napier & Napier, 1967; Eudey, 1987) or *Semnopithecus francoisi poliocephalus* (Corbet & Hill, 1992).

Due to the existence of a similar grey-coloured terminal band on the rump and outer sides of the thighs, Brandon-Jones (1984, 1995) considered *poliocephalus* to be a subspecies of the Hooded leaf monkey *Semnopithecus johnii* from South India, although the two species differ in many other ways. This grouping is largely unaccepted.

Roos *et al.* (2001) found clear differences in genetics and proposed to separate *poliocephalus* as a distinct species with the Chinese White-headed langur *leucocephalus* as subspecies. This classification is also considered by Groves (2001).

3.3.2   Description

The pelage of adult males and females is dark chocolate-brown, not black as in *francoisi* and *laotum*. The head and neck, down to the shoulders, are bright golden-brown to yellowish-white with the head and crest being the most light coloured parts. The species has a 5cm-wide grey band which runs from the thighs to the back, above the root of the tail. This band forms a V-shape, with the tip on the lower back. The hairs of the band are dark brown in color, like the rest of the back, but about 5 to 10mm of the tips are bright silver-grey, giving the band a "frosted" appearance. The feet and hands have a small yellow patch, which is the same color as the shoulders. Females have a pubic patch with pale hairs; they also often have skin patches with irregular pigmentation in the pubic area. Like *francoisi* the Cat Ba langur has very long hair on his "back cape" with a hair length up to 180mm. All bare skin, including the skin of the face, ears, hands and feet, is black.

External measurements (adult)

	n	mm	average	Source
Head/Body length				
male	3	492- 590	537	Brandon-Jones, 1995
female	2	495- 550	523	Brandon-Jones, 1995
Tail length				
male	3	820-872	854	Brandon-Jones, 1995
female	2	820-887	854	Brandon-Jones, 1995

3.3.3   Distribution

The terra typica of *T. poliocephalus poliocephalus* is not specified. The donor of the holotype M. Leger only noted that the species is "not rare in the province of Cai Khin, to the northeast of Tonkin" (Trouessart, 1911). It is not possible to verify the correct identification of the name of the locality and its geographic position. Transcriptions of Vietnamese names often vary, and many different places in the same area have regularly the same name. Additionally, names of places frequently change.

Fooden (1976) mentioned the village Cai Kien (21°19'N, 107°44'E) as a presumed locality. Cai Kien is a small island about 25km south of the Chinese border. However, the island, now called Cai Chien, has no suitable habitat for langurs and did not have such habitat in the past.

The only known locality where the Cat Ba langur occurs is the Island of Cat Ba in Ha Long Bay, close to Hai Phong City.

There is no evidence that the Cat Ba langur ever occurred on the mainland. *T. p. poliocephalus* is closely associated with limestone areas – as are the other northern species of the *francoisi*-group: *T. francoisi* and *T. p. leucocephalus* – and there is no limestone habitat close to Cat Ba on the mainland.

Fossil primates that belong to the colobines are not classified to species level yet, but they are known only from the area where *T. francoisi* occurs (Lang Son Province), and not from the mainland close to the coastline (Nisbett & Ciochon, 1993).

It is highly unlikely that the *poliocephalus* population on Cat Ba Island is a relic population of a formerly widespread species that also occurred on the mainland. The Cat Ba population has experienced a longer development period due to its restriction to a relatively small area, which probably included only the limestone range around Cat Ba (now separate islands) during a time when there was a lower sea level in the Sunda shelf.

3.3.4   Cat Ba langur records

Cat Ba Island (HAI PHONG)
Special-use forest: Partly includes Cat Ba National Park
Cat Ba langur status: Occurrence confirmed, last evidence in 2003 (Stenke, pers comm.)

FZS conducted a first detailed survey for the species from November 1999 until March 2000 and did a first exact census of the population (Nadler & Ha Thang Long, 2000). With the inclusion of sightings prior to the survey (since 1998) in several localities the population was estimated to comprise a total number of 104 to135 individuals (Table 3.3.4-1). Extremly high hunting pressure caused a dramatic decline during and after the survey.

A recent intensive survey conducted as a component of the ZSCSP "Cat Ba National Park Conservation Project" confirmed the existence of only 50 to 60 animals until December 2001 (Table 3.3.4.-2; Stenke, 2001c).

Two langur groups were found on smaller islands. It can be assumed that in the past, the langurs moved from the main island across a mangrove-covered bay. These animals are now isolated due to mangrove destruction.

DISTRIBUTION OF CAT BA LANGUR (*Trachypithecus poliocephalus poliocephalus*) **IN VIETNAM**
BẢN ĐỒ PHÂN BỐ LOÀI VOỌC CÁT BÀ Ở VIỆT NAM

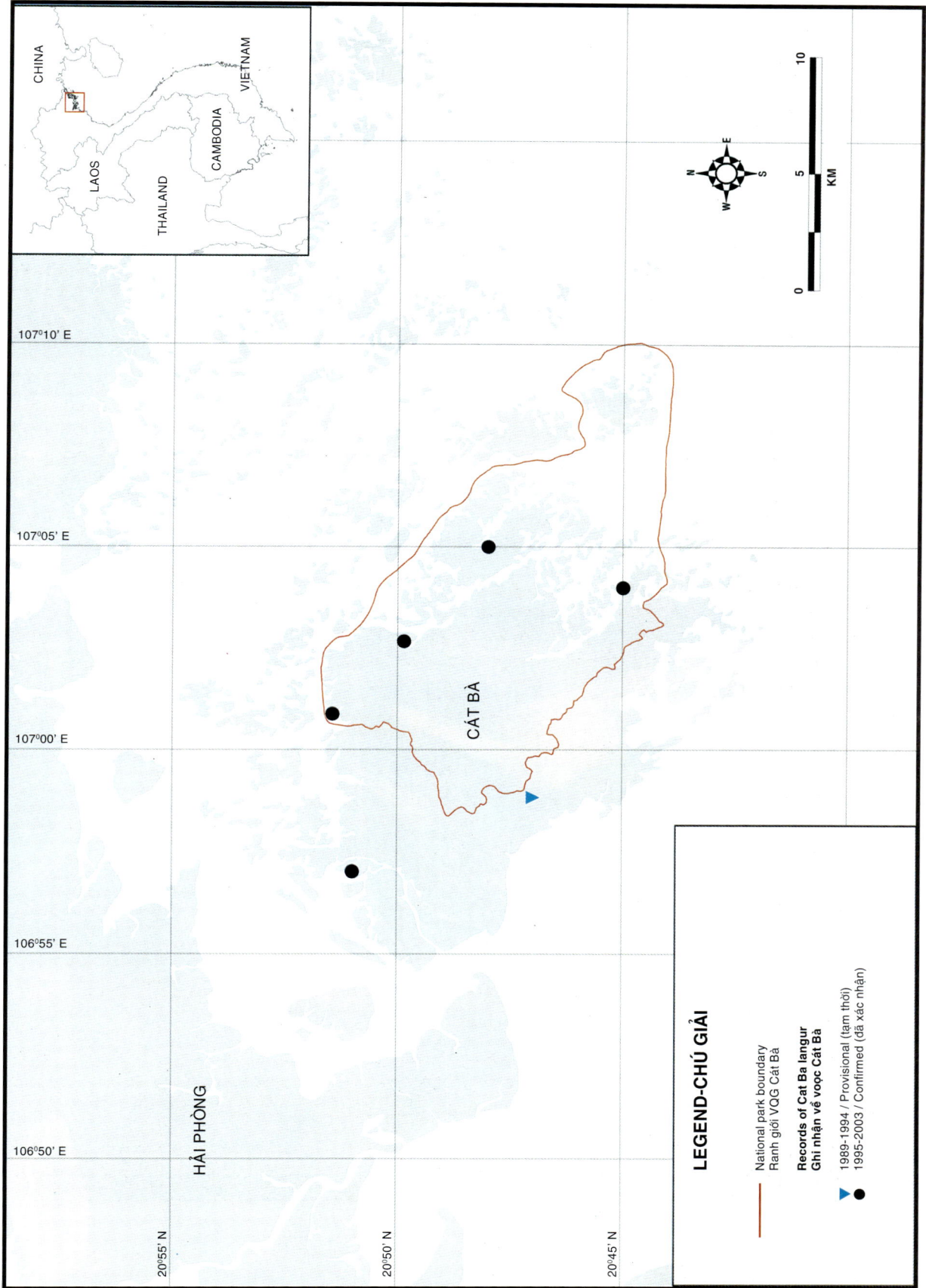

CHINA

VIETNAM

LAOS

THAILAND

CAMBODIA

10

5

KM

0

N E S W

107°10' E

107°05' E

107°00' E

106°55' E

106°50' E

20°55' N

20°50' N

20°45' N

CÁT BÀ

HẢI PHÒNG

LEGEND-CHÚ GIẢI

National park boundary
Ranh giới VQG Cát Bà

Records of Cat Ba langur
Ghi nhận về voọc Cát Bà

▶ 1989-1994 / Provisional (tạm thời)
● 1995-2003 / Confirmed (đã xác nhận)

Table 3.3.4-1
Cat Ba langurs records from 1998 until March 2000 (Nadler & Ha Thang Long, 2000)

Locality	Number of groups	Number of individuals	Remarks
Ang Vem	1	3	Inside Cat Ba NP; probably in contact with Cong Ke group
Cong Ke	1	7	Inside Cat Ba NP; probably in contact with Ang Vem group
Cai Lang Ha	1	10	Inside Cat Ba NP;
Ang Ca - Ang Phay	1	10	Inside Cat Ba NP;
Tung Ba Man - Hen Ca Hong	2	17-19	Inside Cat Ba NP;
Ang Gay - Hang Luon	1	5	Inside Cat Ba NP;
Gio Cung - Ao Ech II	(1)	(7)	Inside Cat Ba NP; erased in January 2000
Van Ta	1	3-10	Inside Cat Ba NP;
Ong Cam	1	4-5	Inside Cat Ba NP; probably in contact with group Ang Le and Uom
Ang Le	1	4-5	Inside Cat Ba NP; Probably in contact with group Ong Cam and Uom
Uom	1	2	Inside Cat Ba NP; Probably in contact with Ong Cam and Ang Le
Tra Bau	1	3-6	Inside Cat Ba NP;
Ang Gian	1	4-5	Inside Cat Ba NP;
Hang Mot - Ang Ke	1	1-3	Inside Cat Ba NP;
Cai So	1	1-4	Outside Cat Ba NP; animals on two small islands
So Tay - Bai Giai	1-2	10-16	Outside Cat Ba NP;
Ang Dai - Ang De Dong Cong	2	14-17	Outside Cat Ba NP;
Lang Cu	1	4-5	Outside Cat Ba NP;
Xom Trong	(1?)	(0-3?)	Outside Cat Ba NP; unconfirmed, erased most probably in 1999
Xuan Dam	(1)	(6)	Outside Cat Ba NP; erased 1998
TOTAL	20-21	104-135	

Table 3.3.4-2
Cat Ba langur records in 2001 (Stenke, 2001c)

Locality	Group Size	Reproduction in 2000/2001	Remarks
Cong Ke 1	6	yes	Inside Cat Ba NP; group isolated on a peninsula, in contact with Cong Ke 2 group; contact with other groups and emigration impossible
Cong Ke 2	2	no	Inside Cat Ba NP; group isolated on a peninsula, in contact with Cong Ke 1 group; contact with other groups and emigration impossible
Cai Lang Ha	4	no	Inside Cat Ba NP; group isolated on a peninsula, contact with other groups and emigration impossible
Peninsula location 1	10(11?)	no	Inside Cat Ba NP; contact with peninsula groups 2 and 3 probable
Peninsula location 2	6	no	Inside Cat Ba NP; contact with peninsula groups 1 and 3 probable
Peninsula location 3	4	no	Inside Cat Ba NP; contact with peninsula groups 1 and 2 probable
Ang Ke	3(?)	yes(?)	Inside Cat Ba NP; unconfirmed
Dong Cong	4	yes	Outside Cat Ba NP; Group isolated on an island, return to main island impossible due to mangrove destruction
Hang Cai 1	6	no	Outside Cat Ba NP; Isolated on a peninsula; contact with Hang Cai 2 group, contact with other groups and emigration impossible
Hang Cai 2	6	no	Outside Cat Ba NP; Isolated on a peninsula; contact with Hang Cai 1 group, contact with other groups and emigration impossible
Cai So	2	no	Outside Cat Ba NP; two females isolated on two small islands; in contact with each other, return to the main island impossible due to mangrove destruction
Ang De - Hien Hao	4-6(?)	yes(?)	Outside Cat Ba NP; Unconfirmed
TOTAL	50-60		

Peninsula locations 1, 2, 3 are equivalent to the locations Tung Ba Man, Ang Gay, Gio Cung, Van Ta, Ong Cam, Ang Le and Uom by Nadler & Ha Thang Long, 2000

3.3.5   Status

The historical (original) density of Cat Ba langurs on the island is unknown. According to older villagers, the Cat Ba langur occurred in high numbers in the past.

By comparing *T. p. poliocephalus* to the relatively well-studied and closely related White-headed langur (*T. p. leucocephalus*), a rough estimation on the historical population density can be obtained. The density of White-headed langurs in optimal habitat is 17.17 individuals per km^2 (Huang Chengming *et al.*, 1998) to 19.71 individuals per km^2 (Li Zhaoyuan *et al.*, in press.). It can be assumed that the entire island of Cat Ba had previously been optimal habitat for the langurs. With a land surface area of 140km^2, Cat Ba could have supported at least 2,400-2,700 individuals.

Most likely the langur population began to decrease in the 1960s, and this decrease was further accelerated by an increase in human settlements in the following decade. However, there is not enough data to accurately assess the demise of the Cat Ba langur population over the years. All previous census data seems generally incorrect. A more accurate and also more impressive number originates from local interviews about hunted langurs: Between 1970 and 1986, some 500 to 800 individuals were killed (Nadler & Ha Thang Long, 2000).

Another rough estimation of the population decline since 1990 pertains to group size, which is independent of the individuals in the whole population. The group size of Cat Ba langurs in healthy habitat is most probably the same or very similar to the group size of the White-headed langurs. In optimal habitat the mean group size of *T. p. leucocephalus* is 9.12 individuals per group (Li Zhaoyuan *et al.*, in press.). For the Cat Ba langur, Nguyen Cu and Nguyen Van Quang (1990) indicate a group size of 9.45 (8.6 to 10.3) individuals per group.

Surprisingly, observations of the White-headed langur show that a drop in the overall population density to 55.5% and a drop in group density to 48.5% do not correspond with the group size (individuals per group) (Li Zhaoyuan *et al.*, in press.). Despite the drastic reduction in population density, for instance, group size has stayed the same. This means that langurs form rather stable social systems and that groups live relatively close to one another. In this way, groups that have experienced a loss of some members will either accept new animals to replenish their numbers or consolidate into new groups. Consequently, group size is not seriously affected as long as individuals interact to form the species' typical social unit (9.12 individuals per group for *leucocephalus*).

Assuming that in 1990, most groups of Cat Ba langurs still maintained a species-specific group size, and considering that *leucocephalus* and *poliocephalus* are very similar in their social behaviour, it can be concluded that most langur groups were still in contact with each other a decade ago.

However, we know there has been a drastic decline in the Cat Ba langur population from 1990 to 2000. There has also been a decrease in group size: 9.45 individuals per group in 1990 to 6.15 individuals per group in 2000 (Nadler & Ha Thang Long, 2000). This indicates already that many of the groups are most probably no longer in contact. An average loss of 3.3 members per group reflects an overall loss of 60 to 70 individuals during the past decade from existing groups. Additionally some groups were totally eradicated. The minimum number of hunted and trapped langurs in this period is 90 to 100. Thus from 1990 to 2000, hunting pressure reduced the Cat Ba langur population by about 50%.

There are reports that during the FZS survey and immediately afterwards several langurs were killed (about 10 are known), and in addition one langur group with 7 individuals was erased (Nadler & Ha Thang Long, 2000). Another group encompassing 10 animals was captured in January 2000 (Baker, 2000). The sole survivor of this group is now kept at the EPRC and involved in a breeding programme. Hunters killed two additional langurs in 2001 and sadly one of these was a pregnant female (Stenke, pers. comm.).

A new census conducted during 2001 as part of the ZSCSP Cat Ba National Park Conservation Project shows a further alarming decline of the population (Stenke, 2001c). The population consists of only 50-60 individuals (Table 3.4.4-2) with a very low reproduction rate.

Compared with the percentage of adult animals in White-headed langur groups – which is 46.2% (Burton *et al.*, 1995) to 55% (Li Zhaoyuan, in press.) – it can be assumed that there are only 30 to 40 adult Cat Ba langurs left.

Besides hunting there is a new threat: the isolation of groups and the loss of panmixia. Without contact with other groups and the exchange of individuals within the population, genetic variability will decrease, and inbreeding will increase. Most of the groups are already in reproductive isolation with no opportunity to exchange or replace group members.

The Cat Ba langur is listed as an **"Critically Endangered"** species with the criteria **CR** A2cd, C2ab, D in the *IUCN Red List of Threatened Species* (Hilton-Taylor, 2000). According to the current situation the Cat Ba langur should be listed with the IUCN criteria CR A1acd, A2cd, B1, B2bde, C2a, D, E. In Vietnam the species is listed as **"Critically Endangered"** (Pham Nhat *et al.*, 1998).

3.3.6   Recommendations for conservation in Vietnam

To preserve the Cat Ba langur from imminent extinction, certain conservation activities must be initiated immediately. According to the results of the surveys, the following measures are urgently recommended:

1. Improvement of forest (habitat) and wildlife protection

2. Management of the remaining population

Some other activities could support the long term survival of the species.

3.3.6.1   Improvement of protection

The main threat to the species is hunting. Hunting is prohibited inside the national park and for protected species in general, but the enforcement of the wildlife protection laws is weak to non-existent.

The following activities should be carried out:

*   The police must inform all villages that no primate can be hunted. Hunting is strictly prohibited.

*   The local authorities and the police must ensure that all guns on the island are registered and must clearly warn people that very high fines will be imposed for unregistered guns (in progress).

*   Current gun owners should be encouraged to sell their guns as an indemnity to the police (in progress).

*   Signs regarding the prohibition of hunting should be displayed in the villages, in Cat Ba town and at all tourist sites, including the harbours on the mainland (Cat Ba, Cat Hai, Hai Phong, Hong Gai) and on all boats.

*   Conservation information should be provided when boat tickets to Cat Ba are issued.

* Written information *from the police* (which has a much stronger influence than the national park) about existing laws, the prohibition of hunting and trading wild animals, and the situation of animals in the wild and their protection should be distributed to all communities, tourist agencies, hotels, and restaurants.

* Informative materials about the laws and regulations for protection and the critical status of wildlife on Cat Ba should be distributed to all schools and schoolchildren (in progress).

* Ranger activities, such as patrols inside the forest and by boat, should be improved and increased (in progress).

* The national park's land and particularly sea borders must be more strictly guarded to prevent illegal entry for hunting, fishing, honey collection, and fuel-wood cutting.

* Laws must be strictly enforced, and the appropriate punishment must be applied for any illegal activity, especially for the hunting and trading of wild animals.

* High-risk key areas, especially certain fjords, should be designated as strictly protected zones. Fishing and any human activities should be forbidden in these fjords.

* Honey collection as a main cause of forest fire and habitat destruction should be strictly prohibited and punished, and any destruction of habitat through fire must also be forbidden and appropriately punished.

* The release of any non-native species, such as Long-tailed macaques *Macaca fascicularis*, must be strictly forbidden and any release of native species must be strictly controlled.

To enforce these activities, a very close cooperation between the national park, the police, the local authorities, and the tourism department on the island and the mainland is necessary. Also necessary are unequivocal regulations regarding protection and responsibilities for the sea area that belong to the national park.

3.3.6.2   Langur population management

Even with strict protection, not all of the langur groups can establish contact with other groups for the purpose of reproduction. Several groups are already in reproductive isolation with no opportunity to exchange or replace group members (Stenke, 2001c).

There are several possibilities to support or restore the panmixia of a population:

In-situ conservation programme
Single animals or small groups with no contact with the rest of the population should be translocated to an area where another group (or groups) exists. This area must be strictly protected against hunting and other human impacts. Local stakeholders must agree that there will be no disturbance and ranger stations and patrols must ensure that there is no human impact.

Keeping animals from moving out of the translocation area can be difficult. To prevent this, the potential translocation site should preferably be an island or a peninsula. A technical device such as an electric fence might be helpful to keep translocated langurs from emigrating, and would constitute an additional means of protection against hunters.

Ex-situ conservation programme
Another strategy to manage and control a small population is to establish a captive breeding population. Confiscated young langurs, injured or handicapped animals should remain in captivity. Groups

caught for a translocation programme could also be divided, and some members used to establish a captive population. This might be a good way to increase the rate of reproduction and ensure genetic diversity (such as through introducing wild surplus males into a breeding programme).

The EPRC successfully keeps and breeds several langur species. The EPRC also has experience in managing groups living in semi-wild conditions in an area surrounded by electrical fencing. Thus the EPRC provides a suitable location to start with the *ex-situ* programme.

The aims of an ex-situ programme should be:

A captive-breeding programme to establish a small captive population.

A release programme to close the gaps between groups in the wild to guarantee genetic flow for the entire wild population.

The *Biodiversity Action Plan for Vietnam* (Government of the Socialist Republic of Vietnam & GEF Project, 1994) lists *ex-situ* conservation projects as part of a coordinated species survival strategy. *Ex-situ* programmes are recommended for taxa that meet three criteria:

* The taxon has a restricted, endemic distribution with a global range of less than 50,000km^2

* The taxon is critically endangered

* In-situ conservation is failing to obstruct population decline.

In the case of the Cat Ba langur, all three criteria are met.

3.3.6.3 National park boundaries

Some langur groups still occur in the northwestern area of Cat Ba Island outside the national park's border. There is no human settlement. The only human impact originates from a few agricultural plots and some cattle grazing in the area's two large valleys. The limestone forest in this area is still in good condition, widely undisturbed. However, habitat destruction is currently increasing steadily through logging.

To prevent human disturbance and forest destruction in this area, the area should be included in the national park. Using the road from Cat Ba town to Gia Luan village as a marker, this northwestern area reaches to about 15km^2 west of the road and 4-5km^2 east of the road. Extending Cat Ba National Park to this area would be an important contribution to protect the langurs and their habitat.

3.3.6.4 Resettlement of the villages inside the park; strict regulations for villages inside the park and along the park border

The villages inside the park, Viet Hai and Khe Sau, as well as Gia Luan, located in the immediate vicinity of the national park, are developing rapidly. The population of Viet Hai has a vast impact on the park. The village has no road access to Cat Ba town, and can only reached by boat. The planned construction of roads through the national park will further accelerate development, agriculture encroachment, and use of forest products.

The heavily cultivated Viet Hai village has already created a deep corridor in the southeastern part of the park. With the enlargement of the village, the whole southeast part of the park, including the large peninsula will be isolated from the central part of the park. At present Viet Hai is inhabited by around 50 families. Many of the houses are simple wood constructions. Viet Hai is still small enough that moving the village is a financially and logistically feasible option.

If resettlement is not possible strict regulations on village extension and use of forest resources should be implemented, including:

* Controlling the human population via a strict limitation on the number of houses.

* Strictly prohibiting the extension of new agricultural areas.

* Strictly prohibiting road-building.

* Strictly prohibiting the use of fuel-wood. (An increase in tourism should not result in an increase in fuel-wood cutting inside the park. Tourists should pay a fee for an alternative fuel source).

3.3.6.5 Economic development for local communities

Improving the living conditions of Cat Ba villagers should go hand-in-hand with reducing negative influences on the environment and the park. No development project should be initiated without preceding environmental impact assessment (EIA), and all projects should have a clear environmental control system. The contents of such projects must be developed in close cooperation with the villagers and local authorities.

A diverse spectrum of projects is conceivable. Some examples are:

* Improving agriculture output by an increase of crop yields instead of extending the land required.

* Intensifying bee-keeping for honey production.

* Developing an ecotourism programme in collaboration with the national park and villages under strict regulations.

3.3.6.6 Ecotourism

Studies and projects regarding the development of tourism (Rauschelbach ed., 1998) should follow, as much as possible, national and international standards such as:

* Ministry of Science, Technology, and Environment, Vietnam: *Enhancing the Implementation of Vietnam's Biodiversity Action Plan* (Hanoi, 1999)

* *Charter for Sustainable Tourism* (World Conference on Sustainable Tourism, Lanzarote, April 1995).

* *Male Declaration on Sustainable Tourism Development* (adopted at the Asia-Pacific Minister's Conference on Tourism and Environment) (Male, February 1997).

* *Berlin Declaration. Biological Diversity and Sustainable Tourism* (International Conference on Biodiversity and Tourism, March 1997).

According to these declarations, some main points are particularly applicable to Cat Ba National Park (Berlin Declaration, 1997):

* Tourism should be restricted, and where necessary prevented, in ecologically and culturally sensitive areas.

* In highly vulnerable areas, nature reserves, and all other protected areas requiring strict protection, tourism activities should be limited to a bearable minimum.

* Tourism in protected areas should be managed to ensure that the objectives of the protected area regime are achieved.

* In all areas where nature is particularly diverse, vulnerable, and attractive, all efforts should be made to meet the requirements of nature protection and biodiversity conservation. Particular attention should be paid to the conservation needs of forest areas, grasslands, freshwater ecosystems, and areas of spectacular beauty.

The document of the Ministry of Science, Technology, and Environment, Vietnam: *Enhancing the Implementation of Vietnam's Biodiversity Action Plan* (Hanoi, 1999) requires:

* The development of guidelines to minimize the impact of tourism in protected areas.

Cat Ba National Park, home of the critically endangered Cat Ba langur, is undoubtedly a very sensitive ecological area. The park's core zone, which holds the remaining primary forest on Cat Ba, is a particularly vulnerable area. As a result, ecotourism in the core zone is not recommended until the national park can be effectively managed and strict regulations can be developed. For example, waste disposal around the park headquarters, ranger stations, and forest campsites is mismanaged and not acceptable for a protected area.

Tourist trails should not enter the core zone. There are many scenic and beautiful views along the park border, such as from the headquarters to Gia Luan or from the headquarters toward Cat Ba town, for tourists to enjoy. Current tourist activities only further affect forest resources and encourage forest destruction (fuel-wood cutting, hunting, wildlife consumption, honey collection, arson, and improper waste disposal). Tourist activities should be restricted to the peripheral zone of the national park. Tourist paths running parallel to the road from Cat Ba, Khe Sau, park headquarters, and to Gia Luan could be very attractive and would eliminate the heavy impact now placed on the core zone of the national park.

The world's rarest primate, the Cat Ba langur cannot serve as a tourist attraction until its numbers increase to a much safer level. Because of the species' current critical situation, this can not be expected for many, many years. The goal for the species should be to re-establish a population that can survive in the long term. The loss of this primate species would be a tremendous loss not only for Vietnam, but also for the whole world. Additionally, the attractiveness of the Cat Ba langur for tourism is low, particularly when compared with other primates (such as chimpanzees, orang-utans, mountain gorillas, proboscis monkeys, Japanese macaques, and golden snub-nosed monkeys). The number of tourists willing to spend money to see this monkey in the wild is not high, so income generated from langur-focused tourism would be low and would clearly not justify the efforts to make this species a tourist attraction. Tourism for this very fragile species would also present a too high risk to the Cat Ba langur's survival.

3.4. Delacour's langur

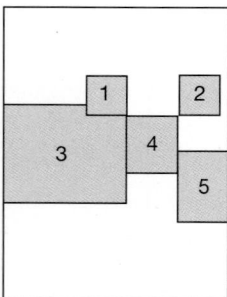

1. Delacour's langur; infant female, one year old. *T. Nadler*

2. Delacour's langur; infant female, six month old. *T. Nadler*

3. Delacour's langur group. *T. Nadler*

4. Delacour's langur; infant male, four weeks old. *T. Nadler*

5. Delacour's langur; ad. male. *T. Nadler*

3.4   Delacour's langur
Trachypithecus delacouri (Osgood, 1932)

3.4.1   Taxonomy

Two animals of this species were collected during an expedition by J. Delacour and W. Lowe close to Hoi Xuan on 15th February, 1930. Osgood (1932) described the animals as a new species *Pithecus delacouri*, Delacour's white-backed langur.

Ellerman & Morrison-Scott (1951) placed this taxon as a subspecies to *T. francoisi*, and most authors follow this classification. Brandon-Jones (1984) reviewed the status and classified the Delacor's langur again as a distinct species. But this view was not widely accepted.

Studies on morphology, behaviour, vocalization and genetics (Nadler, 1994, 1995a, 1996a, 1996b, 1997b; Nadler & Ha Thang Long, 2000; Roos et al., 2001; Nadler & Ha Thang Long, in prep.) clearly show the differences to other taxa at species level (see 2.3). Furthermore, parasitological comparisons support the species' status (Mey, 1994). Most recent publications and reviews have adopted this classification (Brandon-Jones, 1995; Eudey, 1997; Rowe, 1996; Nowak, 1999; Groves, 2001).

3.4.2   Description

The black and white body colouration of the Delacour's langur is unique among the South-East Asian langurs. Apart from the white pubic patch of the females both sexes have the same pelage colour. The black colouration of the upper body is interrupted by a sharply demarcated line in the middle of the back with a white colour pelage between this point and a similar sharply demarcated line just above the knees. As a result the langur looks as if it is wearing a pair of white shorts, indeed the Vietnamese commonly refer to the species as "Vooc Mong Trang" (the langur with white trousers). The light whitish-grey cheek hairs are slightly woolly and longer than in other black Indochinese langurs. The white hairs reach behind the ears where they are tight and form a white patch.

The hair on the head is erected to a crest, the shape is acute and pointed forward. On the back of the head the hair-stroke forms a vertical crest. The hair-stroke on the head is different to the southern *laotum*-species group (Nadler, 1997b).

The long bushy tail differs from all other species of langurs. The hairs are right-angled from the tail so that the tail appears carrot-like with a diameter of approximately 10 cm close to the root.

The only difference in colouration between the sexes is the white pubic patch in females. It is formed by an area of unpigmented skin in front of the callosities with whitish hairs. The pubic patches appear to be irregularly shaped and frequently interspersed with small, sporadic black patches making them individually characteristic. In contrast to *delacouri* and *francoisi*, the *laotum* females have only very small white patches. The pubic patch in *delacouri* and *francoisi*, therefore, appears white with black patches, rather than black with white patches as in *laotum*. The micro-structure of the hair in *delacouri* and *laotum* is also different (Nadler, in prep.).

In some animals the white colouration across the hips occasionally changes to light brown, although more frequently the change in colouration is restricted to the groin area. The white hair on the females' pubic patch can also change colour. The change in colouration is a result of sweat. Females with babies are especially susceptible to having their pubic patch change colour during the summer.

Delacour's langurs are born with a bright yellow-orange pelage. Their colouration is brighter than in *francoisi* and *laotum*. In contrast to these species the bare skin of the face is more yellow, wheras *laotum* is more grey, the eyebrows are also yellow, not black. Therefore, the face has a quite distinctive appearance. The hair colouration changes during the first five weeks to dark orange, whilst the face, hands and feet become black. By four months of age the whole body changes to black and the parts which later change to white become a washed dark grey. After nine months, the body is black, the "trousers" are dark grey, whilst the head turns light brown. The colouration of the head changes slowly, and some light brown hairs remain until the second year. The change in colouration of the "trousers" to the white of the adults takes about three years.

Juveniles have a tassel at the end of the tail. Only at the age of three years the tail changes to the carrot-like shape with a larger diameter close to the root.

External measurements (adult)

	n	mm	average	Source
Head/Body length				
male	5	570-620	595	Nadler unpubl.
	1	580	580	Brandon-Jones, 1995
female	2	570-590	580	Nadler unpubl.
	1	570	570	Brandon-Jones, 1995
Tail length				
male	5	820-880	860	Nadler unpubl.
	1	855	855	Brandon-Jones, 1995
female	4	840-860	855	Nadler unpubl.
	1	840	840	Brandon-Jones, 1995
Weight		kg		
male	9	7,5-10,5	8,6	Nadler unpubl.
female	5	6,2-9,2	7,8	Nadler unpubl.

3.4.3   Distribution

During the decades following the discovery of the species there was only scant information about the existence and distribution of the species. Hanoi University received four animals from a mountain area in Lac Thuy District, Hoa Binh Province between 1960 and 1965 (Fooden 1996). In 1967, a German magazine published a report about Cuc Phuong National Park including a picture of a stuffed Delacour's langur (Stern & Stern, 1967). It was the second published picture of this species after Osgood's first description.

The first sightings of living Delacour's langurs are reported by Ratajszczak *et al.* (1990) to have been in Cuc Phuong National Park in November 1987 (three animals) and 1989 (two animals; Ratajszczak, 1998). In June 1989 the staff of Cuc Phuong National Park made a first video recording of four animals. Adler (1992) reported further sightings in Cuc Phuong 1990 and 1991. Based on local interviews the *Red Data Book of Vietnam* (Ministry of Science, Techology and Environment, 1992) mention eight locations where Delacour's langurs were believed to occur but not all of these occurrences could be confirmed during the following years.

In 1991 the Frankfurt Zoological Society started the Cuc Phuong National Park Conservation Programme. One of the aims of this programme was to investigate the distribution and biology of the species and to improve the protection of its habitat. During the field survey programme many new areas where Delacour's langurs occur were discovered.

The Delacour's langur is endemic to Vietnam and occurs in a very restricted area of Northern Vietnam which comprises about 5000 km^2 between 20°- 21° N and 105°-106° E. The distribution is closely related to the limestone mountain ranges in the provinces of Ninh Binh, Ha Nam, Hoa Binh and Thanh Hoa. Although a smaller limestone mountain ridge continuously stretches west of the mountain range to a large limestone massif north of Son La there is no evidence of Delacour's langurs in this area. The northwestern border of the distribution area is Mai Chau between the Da River in the North and the Ma River in the South.

The Da River seems to form the northern border of the distribution area. North of the Da River there are only two questionable reports about the occurrence of Delacour's langur: Van Chan (=Nghia Lo) in Yen Bai Province and Xuan Son in Phu Tho Province (Ministry of Science, Techology and Environment, 1992).

Further evidence was found in the northwest, in Son La Province (Nadler, 1996b), based on a photograph of two captured Delacour's langurs. The photograph was taken around 1968-1969 and supposedly came from Mai Son District. This district was thoroughly searched for any indication of Delacour's langur during a three week survey in Son La Province in May 1999, however, no information could be collected on Delacour's langur (Baker, 1999). There is also no information from Moc Chau District, Son La Province (Baker, 1999; Luong Van Hao, 2000) and Da Bac District, Hoa Binh Province (Luong Van Hao, 2000).

In the south the distribution area does not show such a clear demarcation line. There are some smaller isolated limestone areas south of the Ma River. The sole area south of Ma River where the occurrence of Delacour's langurs is confirmed is the limestone complex between Lang Chan and Ngoc Lac. However, this population is now probably extirpated. It seems that this species, historically, did not occur south of the Chu River. All information south of the Chu River could not be confirmed during the recent surveys. There is no connection between the limestone mountain ranges in northern Thanh Hoa Province and southern Nghe An Province. The distance beetween these larger limestone blocks is about 100 km.

The southernmost information on the distribution originates from Huong Son District, Ha Tinh Province, and Con Cuong District, Nghe An Province (Ministry of Science, Techology and Environment, 1992). However, interviews from forest protection personnel, locals and hunters in 1997 (Nadler, unpubl.) supplied no evidence to suggest that the species ever occurred in these areas. The records from Quy Chau, Nghe An Province, Thang Phong, (Ministry of Science, Techology and Environment, 1992; Nadler, 1996) and Tinh Gia, Thanh Hoa Province (Nadler, 1996) are based on incorrect information from forest protection personnel and hunters. In June 1995 hunters allegedly kept two Delacour's langurs in the Thang Phong area. One stuffed Delacour's langur and one skin were seen in Tinh Gia. Based on these specimens, information was spread about the occurrence of Delacour's langur in this area. The specimen in Tinh Gia was most probably hunted in the Bim Son area on the border between Ninh Binh and Thanh Hoa Provinces. A five day survey in August 2000 in villages around the limestone area close to Quy Chau supplied no evidence about Delacour's langurs, and they were not known in former times. The species was unknown to former and active hunters (Luong Van Hao, 2000).

The above mentioned occurrence at Thuong Xuan, Thanh Hoa Province must also be regarded as questionable.

Currently there are 19 known locations where Delacour's langurs occur. These localities represent clearly or relatively isolated populations. Only in a few cases can the possibility of interaction between two sub-populations be assumed.

In three locations Delacour's langurs were reported by local people as extinct.

DISTRIBUTION OF DELACOUR'S LANGUR (*Trachypithecus delacouri*) IN VIETNAM
BẢN ĐỒ PHÂN BỐ LOÀI VOỌC MÔNG TRẮNG Ở VIỆT NAM

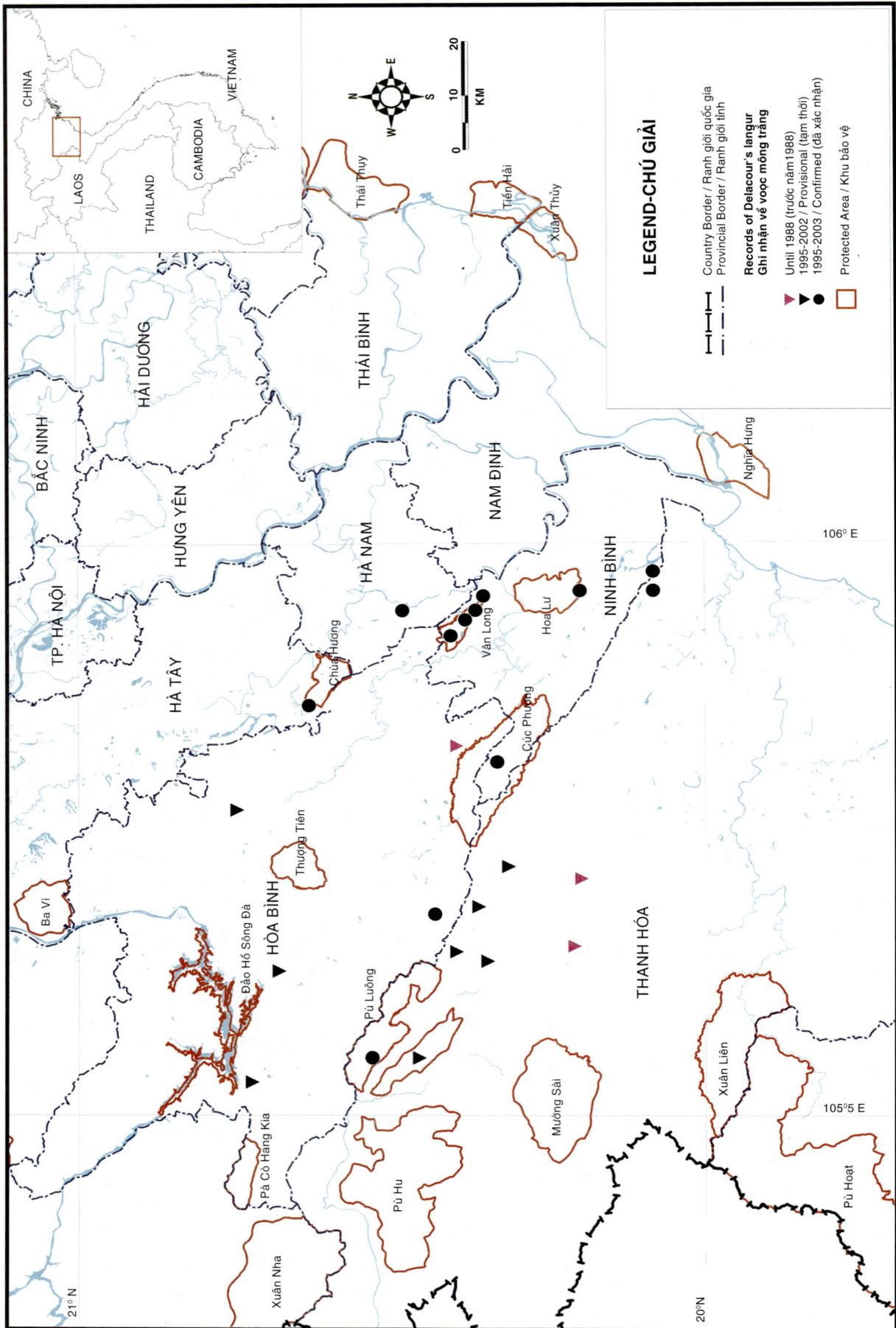

LEGEND-CHÚ GIẢI

Country Border / Ranh giới quốc gia
Provincial Border / Ranh giới tỉnh

Records of Delacour's langur
Ghi nhận về voọc mông trắng

Until 1988 (trước năm1988) / 1995-2002 / Provisional (tạm thời) / 1995-2003 / Confirmed (đã xác nhận)

Protected Area / Khu bảo vệ

3.4.4 Delacour's langur records

Cuc Phuong National Park (NINH BINH, HOA BINH, THANH HOA)
Special use forest: National Park
Delacour's langur status: Occurrence confirmed, last evidence 2003 (Nadler, pers. comm.)

The first ever sighting of a living Delacour's langur was reported from Cuc Phuong in November 1987 (Ratajszczak, 1990, 1998). Further sightings were reported in 1989 (Ratajszczak, 1998) and during the first FZS surveys in November 1990 and in November 1991 (Adler, 1992). Two Delacour's langurs were confiscated from hunters in 1993 and two in 1994. These animals are now housed at the EPRC and are involved in a breeding programme.

Surveys to ascertain the population size were carried out between 1993-1995 (Nadler, 1995, 1996a, 1996b). It was estimated to be 4-5 groups with 20-30 animals. Single sightings and records of faeces under sleeping caves were continously recorded until 2003 (Nadler, unpubl.; Luong Van Hao, 2000a).

Yen Mo limestone mountain range, Yen Mo District (NINH BINH)
Special use forest: None
Delacour's langur status: Provisional occurrence, last report in 1999 (interview)(Luong Van Hao, 1999a)

The occurrence of Delacour's langurs was discovered during the FZS-surveys in 1999.

In 1989 a hunter killed 8 animals of a group close to Mua Thu Village. In 1988-89 a hunter shot three adults and one young animal from a group with about 20 animals on Mot Vai cliff. The rest of the group should still exist. The same hunter shot 2 animals in March 1995 on Da Dung cliff from a group with 8 animals. One of the killed females was pregnant. He had the last sighting in this area in March 1998. In 1989 another hunter shot 6 animals on Da Dung cliff. Another record is of one sighting of 3 animals on Da Dung cliff in August 1999.

One local saw 2 animals on Mot Vai cliff in March 1999. Another local saw a group with 7 in the same location and at the same time of day.

Bim Son mountain, Bim Son District (THANH HOA)
Special use forest: None
Delacour's langur status: Occurrence confirmed, last evidence 1999 (Luong Van Hao, 1999a)

The occurrence of Delacour's langurs was discovered during the FZS-surveys in 1993-1995 (Nadler, 1996b). There were sightings and the number of animals was estimated to be 6-7 individuals. In 1994 one group of six Delacour's langurs was killed by hunters. The entire group was allegedly shot inside a cave; two skins were kept and stuffed (Baker, 1999).

During a survey in April 1999 reported sightings were generally of two adults, and the total number of individuals was estimated to be 2-6 in 1-2 groups (Baker, 1999). During another survey carried out in August 1999 four to five groups were found with a total number of 14-18 animals (Luong Van Hao, 1999a). The groups occurred in 4 locations and it seemed that some groups were probably already isolated.

Hoa Lu Cultural and Historical Site (NINH BINH)

Special use forest: Protected area, cultural and historical site
Delacour's langur status: Occurrence confirmed, last evidence in 1995 (Nadler, 1996b), last report in 1999 (interview) (Luong Van Hao, 1999a)

The occurrence of Delacour's langurs was discovered during the FZS-surveys in 1993-1995 (Nadler, 1996b). There were an estimated 3(-4?) groups with a total number of 10-15 individuals. Based on interviews during a survey in July 1999 the number of Delacour's langurs was estimated to be 3 groups and a total number of 14 individuals (Luong Van Hao, 1999a).

Van Long Nature Reserve (NINH BINH)

Special use forest: Nature reserve
Delacour's langur status: Occurrence confirmed, last evidence 2003 (Nadler, pers. comm.)

The occurrence of Delacour's langurs in this area was discovered during the FZS-surveys in 1993-1995 (Nadler, 1996b). In the overview by Nadler (1996b) the locality and number of animals are included in the locality Da Han, Gia Vien. Another small Delacour's langur population was discovered during a FZS survey in Hang Tranh mountain in 1999. A connection between the three populations is now very unlikely. The areas are separated by a road, agriculture, plantations and increasing numbers of settlers. The newly proposed Van Long Nature Reserve encompasses the three areas where Delacour's langurs occur: Van Long mountain chain, Hang Tranh mountain and Gia Vien limestone area.

Since the FZS survey in the Van Long area in 1993-1995 there have been continuous sightings. A one-month survey in September 1999 in this relatively small area brought an informative overview concerning the abundance and group structure of the Delacour's langur (Luong Van Hao, 1999). During this time 3 groups were recorded, and one or two single animals, altogether 29-31 individuals.

In April 2000 three animals were born and recorded on video. This population comprised 35-40 individuals in 2002 and is the largest presently known.

In June 1997 one Delacour's langur was hunted in the area (La Dang Bat, 1998 and information from FPD Ninh Binh and locals).

The Delacour's langur population in Van Long mountain area was in the past connected with the population on Hang Tranh mountain but it seems unlikely that an exchange of animals is still possible.

The Hang Tranh mountain, a small isolated limestone hill is connected by a 80m-long dam (through a water reservoir) with the Van Long mountain chain. During a one week survey in this area in October 1999 a group of 7 individuals was observed (Ha Thang Long, 1999). Two animals were seen in March 2000, and four in September (Nadler, pers. comm.).

It can be assumed that in the past, interaction between the two areas was possible. However, in 1999 houses were set up on the end of the dam and it seems unlikely that langurs have passed the dam since.

The occurrence of Delacour's langurs in Gia Vien limestone area, now belonging to Van Long Nature Reserve, was discovered during the FZS-surveys in 1993-1995 (Nadler, 1996b). The area has occasionally been visited during the following years, mostly following information from villagers about sightings of Delacour's langurs. There are reports of sightings in many different places but the observations appear to be concentrated in two areas: on the Ba Chon mountain and on the Da Han mountain close to Gia Hoa Village (Nadler, pers. comm.). Three hunted animals were seen in 1993 in Da Han Village. The last information from locals regarding sightings was from January 2002.

Despite the high quantity of information and sightings, it is not possible to give an exact number of the individuals in this area. A rough estimation would be around 20 animals.

Since 1997 the officials of Gia Van commune have tried to develop the Van Long area as a tourist attraction. Visitors can go by boat on the dammed-up stream along the steep limestone outcrops. It is the only place where Delacour's langurs are regularly seen.

Lac Thuy - Kim Bang area, Lac Thuy District (HOA BINH) Kim Bang District (HA NAM)
Special use forest: None
Delacour's langur status: Occurrence confirmed, last evidence 1996 (Nadler, pers. comm.), last report 1999 (interview)(Ha Thang Long, 1999)

The occurrence of Delacour's langurs here is known since 1960 when Hanoi University received 4 specimens from this area: one adult male in May 1960 from Dong Tam, one juvenile male in July 1960 from Phu Thanh, one juvenile male in February 1961 from Lac Thuy and one adult female in May 1965 from Lac Long (Fooden, 1996).

During the FZS-surveys in 1993-1995 there were sightings of 3 groups of Delacour's langurs (with about 14-16 individuals). Based on information from hunters and locals, Nadler (1996b) estimated the number of individuals in this area to be about 30-40. Three animals were observed in November 1996 (Nadler, unpubl.). During a FZS-survey in October/November 1999 (Ha Thang Long, 1999) hunters and locals confirmed the existence of one group with about 20 individuals.

The hunting pressure in this area is very high and some hunters specialize in hunting langurs. On the basis of hunter information the number of hunted Delacour's langurs prior to 1996 had been around 20 per year (Nadler, unpubl.).

One Delacour's langur was found in Bao Sao village in October 1993 (Kim Bang District) in a cooking pot in the process of preparing "balm" (Nadler, unpubl.). One langur group of 11 individuals was observed in November 1993, which had apparently declined to 3 animals by 1996, according to information from hunters. Another group of 9 animals was caught in a sleeping cave in July 1995 and all animals were killed. Three animals from this group were seen as stuffed specimens in the homes of hunters in Dong Tam Village, Lac Thuy District (Nadler, unpubl.). A villager from Dong Tam brought an approximately three month old Delacour's langur to the EPRC in April 1995. The mother was hunted in this area. This langur is now involved in a breeding programme at EPRC.

Huong Son mountain, My Duc District (HA TAY), Lac Thuy District (HOA BINH)
Special use forest: Partly included in Huong Son Cultural and Historical Site
Delacour's langur status: Occurrence confirmed, last evidence in 2000 (Luong Van Hao, 2000a)

The first information about the occurrence of Delacour's langur was collected during the FZS survey in 1996 (Nadler, unpubl.). During the FZS survey conducted in December 1999/January 2000 locals and hunters named six localties where Delacour's langurs occurred.

Close to Yen Vi village one skull was found in the forest. Two Delacour's langur tails are kept by hunters. One of the animals was hunted in 1995 (locality unknown) and the second one was hunted close to Yen Vi. A local in Ai Lang/An Phu specialized in the production of "monkey balm" and purchased around 100 Delacour's langurs between 1990-99, 39 from this area and the other ones from the adjacent provinces Ninh Binh and Nam Ha.

One animal of a group of 5 individuals (reported by locals) was seen during the survey in January 2000 (Luong Van Hao, 2000a).

Roc mountain, Kim Boi District (HOA BINH)

Special use forest: None
Delacour's langur status: Provisional occurrence, last report 2000 (interview, specimen) (Luong Van Hao, 2000a)

Indications of the occurrence of the species in this area were found during the FZS survey in 2000. There were two places identified by locals where Delacour's langurs were believed to occur: Da Roc and Da Chi. In Da Chi the langurs were extirpated around 1960. This population had been about 10 individuals. Two hunters killed 12 individuals in both areas. In Da Roc now only two animals are left.

A hunter in Xom Hoi village keeps a tail of a Delacour's langur hunted in 1985.

According to information from hunters this area is the northernmost part of the range of Delacour's langurs.

Phu Vinh mountainous area, Tan Lac District (HOA BINH)

Special use forest: None
Delacour's langur status: Provisional occurrence, last report 1999 (interview) (Luong Van Hao, 1999b)

Delacour's langurs were found to occur here during the FSZ-survey in 1999. Based on information from locals, two groups of langurs with a total of 15 animals occur close to the Song Da River / Hoa Binh Lake. Around 1989, 4 groups were hunted (Luong Van Hao, 1999b).

Mai Chau mountainous area, Mai Chau District (HOA BINH)

Special use forest: None
Delacour's langur status: Provisional occurrence, last report 2000 (interview)(Luong Van Hao, 2000a)

The FZS survey in 2000 found langurs to occur in this area. Based on interviews there were two areas where Delacour's langurs occurred: Pom Xom mountain and Tan Mai - Ba Khan mountain. The population on Pom Xom mountain, with 8 groups and 40 individuals, was extirpated in 1980.

The remaining population on Tan Mai - Ba Khan mountain is about 3 groups with a total of 15 individuals.

Pu Luong Nature Reserve (THANH HOA)

Special use forest: Nature reserve
Delacour's langur status: Occurrence confirmed, last evidence 1999 (Baker, 1999)

The Pu Luong Nature Reserve is separated into two parts: the smaller south-western part includes Hoi Xuan - the terra typica of this species - Phu Nghiem, Thanh Xuan, Phu Xuan and Phu Le and the larger north-eastern part includes Lung Cao, Co Lung, Thanh Son and a part of Phu Le.

The occurrence of Delacour's langur is noted in a report on the flora and fauna in Pu Luong (Le, 1997). The author of this report estimated that 50 individuals live in Pu Luong. A FZS survey was conducted in 1999 to verify the information.

Delacour's langur were described in the south-western area by locals from Hoi Xuan, but this information was vague. One hunter recalled killing two Delacour's langurs around 1978. Ratajszczak *et al.* (1990) figure a photo from a Delacour's langur skin which was offered for sale at NAFORIMEX's Thanh Hoa office in September 1989. The provincial representative of the company claimed that the animal was captured in Quan Hoa district. The most recent sighting was several years previous ago. In Tan Son Village (Thanh Xuan), the only village inside the Pu Luong reserve boundary, Delacour's langur was known but the last sighting dated back 10 years, although local reports revealed that Delacour's langurs are still present around Mt. Pha May. One individual was allegedly killed in 1998. Reports claimed that Delcaour's langurs were abundant 10 to 20 years ago (groups of 20 to 30 individuals were common), but are now very rare.

In Phu Xuan Village the Delacour's langur was well known but the most recent sighting was about 8 years prior to the survey. One skin was kept by a hunter. A hunter in Pu Nhiem village reported that he had hunted one Delacour's langur in about 1989 but this species was now extirpated in this area and there were no more sightings from locals.

The estimated number in the south-west part of Pu Luong Nature Reserve is 2-3 groups with about 10 animals (Baker, 1999).

The north-east part of the Pu Luong Nature Reserve has more remote areas with more intact habitat and the best forest coverage. The largest numbers of Delacour's langurs in Pu Luong probably live close to Ho and Lang Hang (Lung Cao Village). During the time of the survey in February 1999 a hunter killed one Delacour's langur from a group with 7 animals close to Lang Hang village.

Close to Co Lung village, locals reported a sighting of a group with 5-6 individuals during the survey. The estimated number in the north-east part of Pu Luong Nature Reserve is 5-6 groups with about 30-35 individuals (Baker, 1999).

Ngoc Son mountainous area, Lac Son and Tan Lac Districts (HOA BINH)
Special use forest: Partly proposed nature reserve
Delacour's langur status: Occurrence confirmed, last evidence 1999 (Luong Van Hao, 1999b)

Delacour's langurs were discovered in this area by a FZS survey in 1999.

There are three localities (Bo Village, Ngo Luong Village; Ngoc Son Village) where Delacour's langurs occur.

Close to Bo Village one group of 5 to 7 animals exists. The last sighting from villagers was reported in January 1999. Another group with 10 individuals was extirpated in 1990 in this area.

Locals reported one group with 20 animals close to Ngoc Son Village in 1998. One animal was killed in December 1998 and another one in May 1999. More langurs probably were hunted during this time because, when the group was last observed in June 1999, it consisted only of 4 animals.

Northern Ba Thuoc mountainous area, Ba Thuoc District (THANH HOA)
Special use forest: None
Delacour's langur status: Provisional occurrence, last report June 1999 (interview)(Baker, 1999)

Delacour's langurs were discovered here during the FZS survey in 1999. Direct sightings of Delacour's langurs were made in one area.

In Ai Thuong, local reports claimed that groups of Delacour's langurs used to come to the cliffs just above the village, most probably one group with 5 individuals. Since 1994, the village of Ai Thuong, in cooperation with Luong Noi Village and a district organization, have been protecting this forest area in a local programme. However, it is not clear how effective this protection has been for the existence of the langurs.

Thiet Ong mountain, Ba Thuoc District (THANH HOA)
Special use forest: None
Delacour's langur status: Provisional occurrence, last report 1999 (Baker, 1999)

Delacour's langurs were discovered to occur here during the FZS survey in 1999.

It is a small and isolated limestone complex surrounded by the Ma River. The occurrence of Delacour's langurs was reported by FPD rangers and locals. One langur was hunted in 1990 and the hunter still kept a part of the white fur of the hind quarter. The most recent sighting of one group with 11 individuals were as made in April 1999.

Nui Boi Yao mountainous area, Lac Son District (HOA BINH), Ba Thuoc District (THANH HOA)
Special use forest: None
Delacour's langur status: Provisional occurrence, last report 1999 (Luong Van Hao, 1999b)

Delacour's langurs were discovered in this area during the FZS survey in 1999. The area included 4 localities (Ngoc Lau Village; Khu Thuong, Tu Do Village and Lang Dam, Luong Noi Village).

According to local interviews 1 group with 5-7 individuals exists close to the boundary of Thach Lam Village (Thach Tuong mountainous area). Another group close to Khu Thuong with 2-3 individuals was extirpated around 1990.

Around Tu Do Village there were 4 groups with around 30 individuals until 1994. Now the langurs are extirpated in this area. In Lang Dam, Luong Noi Village the langurs are well known. The most recent sightings were made in 1999. Two sleeping sites were identified and checked, both were possibly used by the same group. Fresh faeces were found at both sites.

Thach Tuong mountainous area, Thach Thanh District (THANH HOA)
Special use forest: None
Delacour's langur status: Provisional occurrence, last report 1999 (interview) (Baker, 1999)

Delacour's langurs were discovered to occur here during the FZS survey in 1999.

On the north side of Thach Tuong, the forest is connected with the westernmost tip of Cuc Phuong National Park and is only separated by a dirt road along a small stream. But in the adjacent part of the national park the Delacour's langurs are probably already extirpated (Luong Van Hao, 2000b).

A hunter reported one group of Delacour's langurs with 5 individuals. He knew the sleeping cave and had tried to catch this group several times.

Locations where Delacour's langurs are believed to be extinct

Nui Ke mountainous area and vicinity, Yen Thuy and Lac Son Districts (HOA BINH)

This area has very small and isolated limestone outcrops. Locals reported about 8 isolated limestone hills where Delacour's langur occurred in the past (Nui Ke, Quen Liep, Nui Chong Chang, Nui Hang Nga, Nui Hang Gio, Nui Quang Khe, Nui Thong Bo and Nui Dam Hong).

The Delacour's langur were extirpated around 1980. The population in this area was once about 70 individuals according to local information (Le Thien Duc, 2002).

Lang Chanh mountainous area, Lang Chanh, Ngoc Lac and Thuong Xuan Districts (THANH HOA)

Dao Van Tien (1985) listed one Delacour's langur collected in Tan Phuc in March 1964. The information given by Ratajszczak (1990) regarding the occurrence of the species in Thuong Xuan District probably concerned the same limestone range in the northern part of the district.

During the survey conducted by the FZS in 1994 there was one sighting of two animals and, based on information from hunters, it was estimated that 10-15 individuals lived in this area (Nadler, 1996b).

In a further survey in 2000 the hunters and local people interviewed verified that Delacour's langurs were already extirpated (Luong Van Hao, 2000c).

Mountainous area in vicinity of Cam Thach village, Cam Thuy District (THANH HOA)

During the survey conducted by the FZS in 2000 the hunters and local people interviewed verified that Delacour's langurs were extirpated in 1987 and the last animals were hunted (Luong Van Hao, 2000c).

Discussed records

Nghia Lo, Nghia Lo District (YEN BAI) 21°36' N / 104°31' E

The locality is mentioned in the *Red Data Book* (Ministry of Science, Techonology and Environment, 1992) but during a survey conducted by the FZS in 1995 there was no information about the occurrence of Delacour's langurs (Nadler, 1996b).

Son La mountainous area, Muong La, Son La, Mai Son, Thuan Chau and Moc Chau Districts (SON LA) 20°30' - 21°30' N / 103°45' - 105°00' E

One report suggesting the occurrence of Delacour's langurs (Nadler, 1996b) and one photograph of two caught Delacour's langurs allegedly from Son La limestone area were the basis to survey this large area. If Delacour's langurs occur - or occurred - in Son La, this province would represent the north-westernmost area of the species' range. The langurs in the photograph, circa 1968/69, were supposedly from Mai Son District, thus an emphasis was placed on thoroughly surveying this district.

Every district FPD in Son La Province was interviewed and local FPD officials had never heard of Delacour's langurs occuring in Son La Province.

After a three-week survey conducted by FZS in 1999 no information could be collected on Delacour's langur (Baker, 1999). Another survey to determine the north-western extent of the species distribution was conducted in 2000 in Moc Chau district and also brought no evidence of their occurrence (Luong Van Hao, 2000a).

Xuan Son Nature Reserve (PHU THO) 21°05' N - 21°11' N / 104°50' E - 104°58' E

The Delacour's langur is mentioned in the proposal (Le Xuan Canh, pers. comm.) for the establishment of the protected area but there is no verification of the occurrence of the species.

Tinh Gia, Tinh Gia District (THANH HOA) 19°27' N / 105°42' E

A Delacour's langur's tail at a hunter's house in Tinh Gia and the mistaken information that the animal was hunted in a small limestone range nearby was the reason for the assumed occurrence in this area (Nadler, 1996b). There was no reliable evidence that Delacour's langurs ever had occurred in this area.

Tan Phong, Quy Chau District (NGHE AN) 19°38'N / 105°19'E

Information about the occurrence of Delacour's langurs was based on two caught animals, seen in 1994 as stuffed specimens (Nadler, unpubl.). But further information indicates that these animals are from another (northern) area and there is no evidence to support their occurrence in this locality.

Quy Chau, Quy Chau District (NGHE AN) 19°33' N / 105°05' E

The occurrence of Delacour's langurs is mentioned in the *Red Data Book* (Ministry of Science, Technology and Environment, 1992). The confirmation by Nadler (1996b) is based on incorrect information. One survey in this area conducted by the FZS in August 2000 found no evidence to support the occurrence of this species (Luong Van Hao, 2000c).

Con Cuong, Con Cuong District (NGHE AN) 19°01' N / 104°58' E

The occurrence of Delacour's langurs is mentioned in the *Red Data Book* (Ministry of Science, Technology and Environment, 1992).

Interviews during a short survey in 1997 in a limestone area south-east of Con Cuong conducted by the FZS supplied no information about the occurrence of Delacour's langurs (Nadler, unpubl.).

Huong Son, Huong Son District (HA TINH) 18°31' N / 105°26' E

The occurrence of Delacour's langurs is mentioned in the *Red Data Book* (Ministry of Science, Technology and Environment, 1992). In this area there is no limestone mountain range and it is very unlikely that a population of Delacour's langurs occurs so far from the distribution area and in a different habitat.

3.4.5   Status

Table 3.4.5-1 gives an overview of the estimated groups and individuals in the 19 isolated sub-populations of Delacour's langur. The total number is 49-53 groups with a total of 270-302 individuals for the whole species.

11 additional animals are kept at the Endangered Primate Rescue Center; six of these were born there.

In average there are 5-6 indivuals/group. Compared to similar species of Indochinese langurs (*T. francoisi, laotum, poliocephalus, leucocephalus*) this indicates already a reduction in the population density. The average number of individuals/group in a normal population density is around 9 (compilation see Nadler & Ha Thang Long, 2000, see also 3.3.5).

The abundance of the sub-populations indicates a very high number of small isolated groups (table 3.4.5-2). 20% of all existing Delacour's langurs occur in isolated populations with a maximum of 10 animals. Such small populations are extremly sensitive and vulnerable. The loss of the important male for reproduction - mostly only one per group - causes the annihilation of the whole sub-population. The long term existence of these sub-populations is also doubtful for genetic reasons.

Without management and strict regulations the loss of these sub-populations and consequently of 20% of the population of this species is foreseeable.

55 % of the population, sub-populations with 11-30 animals are also particularly vulnerable if it is not possible to stop the hunting. Unfortunately, it can be assumed that there will be no effective control of hunting within the immediate future.

Only two solitary sub-populations comprising 30-35 individuals, the necessary to survive, are known. Luckily one sub-population exists in Van Long Nature Reserve. This area also has an advantageous geomorphological structure which makes access more difficult and can be easily controlled. However, the genetic structure of this sub-population is unknown and studies are necessary to assess the genetic diversity as a basis for long term survival.

The most important, and for some sub-populations, the only reason for the decline is hunting pressure. During the FZS surveys an attempt was made to record all hunted Delacour's langurs (Table 3.4.5-3). Certainly this list is not complete. Some hunters were not found, some did not remember the exact numbers and some were afraid to give correct information. But these numbers are alarming enough. The recorded total number of hunted animals over a period of 10 years is 316. This equates with an annual loss of more than 30 individuals. The numbers hunted in the late 1980's could not be calculated exactly but it was estimated at around 50 or more animals per year.

Since 1996, there has been a clear reduction in hunting levels. It has been influenced by the break down of some sub-populations and the disappearance of langurs in some areas. Possibly it has also been influenced to a small degree by the improvement of ranger activities and increased law enforcement.

Based only on the known hunted langurs, the population has declined by 50-55% over the last 10 years. Normally, the decline of a population decreases with the reduction of the absolute numbers of individuals. The last single individuals are not so easily hunted anymore. The Delacour's langur has not had a closed distribution area for a long time. A single group as a reproductive unit is much more vulnerable. In some cases where a sub-population is only represented by a small number of animals, the capturing of one group can result in the collapsing of the whole sub-population. If only one group exists in a sub-population the sub-population is easily extirpated.

Table 3.4.5-1
Number of groups and individuals of Delacour's langurs in the sub-populations

No.	Locality	Groups	Animals
1.	Cuc Phuong National Park	4-5	20-25
2.	Yen Mo limestone mountain range	2	10
3.	Bim Son mountain	4-5	14-18
4.	Hoa Lu - Tam Coc - Bich Dong	3	14
5.	Van Long Nature Reserve (mountain chain)	4	35-40
6.	Van Long Nature Reserve (Hang Tranh mountain)	1	7
7.	Van Long Nature Reserve (Gia Vien limestone area)	1	4-5
8.	Lac Thuy - Kim Bang area	1-2	20
9.	Huong Son mountain	6	27
10.	Roc mountain	1	2
11.	Phu Vinh mountainous area	2	15
12.	Mai Chau mountainous area	3	15
13.	Pu Luong Nature Reserve (North-East part)	5-7	30-35
14.	Pu Luong Nature Reserve (South-West part)	2	10
15.	Ngoc Son mountainous area	2	9-11
16.	Northern Ba Thuoc mountainous area	1	5
17.	Thiet Ong mountain	1	11
18.	Nui Boi Yao mountainous area	4	17-27
19.	Thach Thanh District	1	5
Total		49-53	270-302

Table 3.4.5-2
Classification and numbers of sub-populations on Delacour's langurs

Number of indivuals in one sub-population	1-10	11-20	21-30	31-40
number of sub-populations	8	6	3	2
number of individuals in the sub-populations	52-55	89-93	64-79	65-75
percent of individuals of the total population in the sub-populations	20	30	25	25

Table 3.4.5-3
Hunted Delacour's langurs

Year	1990	1991	1992	1993	1994	1995	1996	1997	1998	1999
hunted individuals	50	36	38	44	46	49	12	13	14	14

The global conservation status of the Delacour's langur in the *IUCN Red List of Threatened Species* (Hilton-Taylor, 2000) is **"Critically Endangered" CR** A1d, C2a. The species is also considered as **"Critically Endangered"** in Vietnam (Pham Nhat *et al.*, 1998).

With the results of the surveys the species should be listed in the category **"Critically Endangered"** with the criteria **CR** A1acd, A2cd, C2a.

3.4.6   Recommendations for conservation in Vietnam

With the Delacour's langur being a species endemic to Vietnam the responsibility for the conservation and preservation of this species rests upon Vietnam's government as a national task. The status of the species in the wild and the alarming levels of hunting pressure show that the species is on the brink of extinction. Effective on-the-ground conservation measures are an immediate necessity.

3.4.6.1   Improvement of protection and law enforcement

One obstacle to improving the protection of this species is the fragmented distribution. Most of the areas where sub-populations of Delacour's langurs occur are surrounded by areas with dense human population. Only six areas have a protection (or proposed protection) status and are under the direct control of the forest administration.

 The controls in and around protected areas must be intensified. Access to protected areas is very often regularly controlled only in the most frequented parts. But animals and hunters are normally found in the most remote areas.

The control of firearms and the trade in animals and animal parts, also stuffed animals, must be strictly enforced. Wildlife restaurants and middle-man traders for wildlife are still common around protected areas.

The urgent necessity for the protection of all sub-populations - especially outside protected areas - exceeds the capacity of the forest protection departments and forest ranger stations at present. An improvement in protection is not possible without strong support from local authorities. The local authorities must be involved in any protection project. To motivate the local authorities is a difficult task because there is rarely any level of awareness about conservation needs.

The hunting pressure cannot be reduced in the short term with a national gun ban or gun control. Enforcement must take place on a local level if it is to be effective. Hunters and gun owners are usually well known by the locals. Delacour's langurs can only be hunted by hunters which are familiar with the locality. The common explanantion that the animals are hunted by strangers to the locality is an excuse for violation of the law.

A broad education campaign for the public across the entire distribution area which is powerful enough to improve protection and to reduce hunting pressure cannot be realized. Regarding the critical conditions for the species' survival it seems much more important to involve the local authorities in an administrative role in the national task of preserving an endemic species.

In this special case measures must be taken, in particular to involve other governmental authorities besides the forest protection authorities. An efficient tool could be a direct letter from the government - Minister or Prime Minister - to the local authorities located around important distribution areas demanding the implementation of measures to protect the species and to monitor those populations.

The strict enforcement of the legally imposed fines also needs additional attention. In some remote areas, where hunting and the possession of firearms are common, the violations of the law are not notified and not punished.

It is equally important to prosecute the illegal possession of firearms and the trade of endangered species which does not directly involve the hunters. Endangered animals are often offered in public. For example, there were three Delacour's langurs offered in a shop along Road No.1 in Tam Diep (Ninh Binh Province) for a long time and there was no prosecution by the responsible forest protection authorities, despite the fact that they were informed.

Violations of the forest protection laws and the punishment should be made public as well.

3.4.6.2   Recommendations for some special areas

Cuc Phuong National Park (NINH BINH, HOA BINH, THANH HOA)
Pu Luong Nature Reserve (THANH HOA)
Ngoc Son Nature Reserve (HOA BINH)
Nui Boi Yao mountainous area (Lac Son District, HOA BINH, Ba Thuoc District, THANH HOA)
Thach Tuong mountainous area (Thach Thanh District, THANH HOA)
These five areas mentioned above are involved in a planned GEF-project on limestone protection. Within this framework a protection programme for the Delacour's langur should be developed.

Hoa Lu Cultural and Site (Hoa Lu and Tam Diep Districts, NINH BINH)
The protected area is a popular place for bird hunting and orchid collection. These activities should be stopped to reduce the impact on the forest. The famous tourist place should be used to educate people about the need to protect forest resources and not as a market place to purchase animals and plants collected in a protected area.

Van Long Nature Reserve (NINH BINH)
Van Long Nature Reserve is one of the most important areas for the future existence of the species, as it is home to the largest sub-population. Therefore, special attention must be given to its protection. There is an interest by local authorities and the commune to develop the tourism potential of the area, with small boat trips along the dammed stream by the steep limestone outcrops where it is possible to watch the langurs in the early morning. It is a pleasant activity for tourists and could be beneficial for the animals as well.

Thus, efforts should be made to improve the direct protection of the species, and all hunting must be stopped. However, it is also necessary to improve habitat protection, to stop the intensive fuelwood cutting and limestone quarrying and drastically reduce the number of domestic goats. The number of domestic goats in this area is far too high in terms of ecological sustainability. Currently there are about 1,800-2,000 goats grazing in the area. They destroy a considerable amount of vegetation and compete with the langurs for food sources. The resulting disturbance through permanent human presence is also very significant.

For the long term survival of this sub-population it would also be important to create a connection to Hang Tranh Mountain and the Gia Vien limestone area which are inhabited by other sub-populations of Delacour's langurs. The connection could be created by reforesting a corridor with trees which the Delacour's langur feeds on. Van Long and Hang Tranh are connected by a dam across a lake. Hang Tranh and the Gia Vien limestone area (Ba Chon Mountain) are separated by a small road and an eucalyptus plantation which the Delacour's langurs avoid crossing. According to locals, animals were crossing between the two areas before the road was upgraded and before there was a plantation.

Huong Son Mountain, My Duc District (HA TAY), Lac Thuy District (HOA BINH)
The protected area as a cultural and historical site with the Perfume Pagoda is one of the most famous tourist places in Northern Vietnam, receiving tens of thousands of tourists every year. The demand for wildlife meat is very high. Wildlife restaurants should be strictly controlled and an education / awareness campaign against wildlife consumption and for nature conservation promoted.

Ngoc Son mountainous area, Lac Son and Tan Lac Districts (HOA BINH)
In the past, this area, which includes a proposed nature reserve, was quite remote. With the recently constructed new road there is now much better access, and higher levels of human impact - including hunting - can be expected. Local authorities should be called upon to support the protection activities

and to follow laws and regulations. The area should be designated as protected area or as a protected forest corridor that links Pu Luong Nature Reserve and Cuc Phuong National Park.

Northeastern Ba Thuoc mountainous area, Ba Thuoc District (THANH HOA)
The efficiency of the district organised protection programme for this area should be checked and the possibilities to improve the protection with the support of local people should be assessed.

3.4.6.3   Protected area system review

Delacour's langurs were found in 19 areas of which 5 have a protection status (Cuc Phuong National Park, Hoa Lu Cultural and Historical Site, Huong Son Cultural and Historical Site, Pu Luong Nature Reserve, Van Long Nature Reserve) and one is proposed as a protected area (Ngoc Son mountainous area).

With the present high hunting pressure, even in protected areas, there is the paradoxical situation that the animals in unprotected areas are often safer than in protected areas. The much better forest cover in protected areas facilitates hunting (such as Cuc Phuong National Park, Pu Luong Nature Reserve). Steep limestone cliffs with degraded forest, bare slopes and shrub vegetation are safer for the langurs because it is more difficult for hunters to approach the animals.

Such comparably degraded areas - for example Van Long Nature Reserve - can also be more easily patrolled and protected. As long as sufficient food tree species grow in such an area, however crippled, and as long as there are adequate limestone cliffs, rocks and caves it can be a suitable habitat for Delacour's langurs.

Most of these areas are not used by local people - except occasionally for fuelwood cutting or limestone quarrying. Therefore, it should be determined which of the degraded areas with a relic population could be a special "Species/Habitat Conservation Area" following the newly proposed classification of protected areas in Vietnam (Decree No. 08 /2001/QD-TTg).

An area could also be declared as "Species/Habitat Conservation Area" if it provides suitable habitats but has no resident Delacour's langur population. Groups from threatened and unsafe areas (for example, through limestone quarrying for the cement industry) could be translocated into such an area. Such "Species/Habitat Conservation Areas" must be of course guarded and controlled. However, there are suitable limestone outcrops surrounded by agricultural land which are relatively easy to protect.

Within the GEF-project for the Cuc Phuong - Pu Luong limestone area the establishment of two additional "Species/Habitat Conservation Areas" should be considered. These areas should cover Nui Boi Yao mountainous area and Thach Tuong mountainous area to create a larger connected area for this species and to involve all marginal groups.

3.4.6.4   Population management

At present it is very questionable whether the small isolated sub-populations of Delacour's langurs will survive unless hunting pressure is reduced and a larger area for their conservation is established in order to guarantee genetic flow between the currently isolated sub-populations.

However, some isolated groups have absolutely no chance of surviving. Genetic isolation and inbreeding is not the highest threat because, before it will have any effects on the group, it will no longer exist. Hunting, disease, accidents or the natural death of an individual important for reproduction are fatal incidents for such small sub-populations. There is no way to replace the lost individual.

To prevent the collapse of the sub-populations and consequently the extinction of the species, conservation measures should be taken immediately. Several conservation strategies should be applied simultaneously:

*   Larger sub-populations which seem to be able to survive in the wild should be monitored and the protection should be improved as much as possible.

*   Small and threatened sub-populations and single groups which are doomed to be extinct should be assigned for translocation. Translocation is a useful conservation technique and there are positive examples for several primate species (Koontz, 1997; Oates, 1999).

       "Translocation is especially appropriate where a population of animals has been greatly reduced or eliminated from an area of suitable habitat, but where the factors that led to the population decline no longer operate; in this case, animals may be translocated from a similar area nearby with a healthy population. Alternatively, animals may be moved from an area where they are under severe threat to an area of lower threat. In either case the animals are not held for an extended period in captivity, and unlike animals bred in a zoo or other long-term facility, they are fully adapted to a wild environment" (Oates 1999).

*   Another method is *ex situ* conservation - to keep animals in captivity. The experience of many years at the EPRC shows that is possible to keep this species successfully in captivity (Nadler, 1999). The important factors for successful keeping are:

1.   Maintaining adequate climate conditions comparable to the climate in the distribution area for the species

2.   Diet composition as similar as possible to the natural food

3.   Housing the animals in social structures comparable to wild groups

The offspring of a small captive population can be used to stabilize existing sub-populations (reinforcement, IUCN 1998) or to re-establish a new sub-population in an area where the species has been extirpated (re-introduction, IUCN 1998).

A important model for a medium or long term conservation strategy is the establishment of semi-wild areas. Keeping animals in a semi-wild area combines the advantages of wild conditions and captivity. The enclosure should be large enough so that animals have natural conditions, including species typical habitat, food, and possibilities to develop normal social intraspecific behaviour. But the area should be small enough to control, to manage and treat the animals and to avoid all illicit access.

The EPRC has set up two semi-wild enclosures surrounded with electrical fencing. In one area one group of Hatinh langurs has been kept successfully and without any problems for more than four years. The second area is large enough to keep more than one group. These areas are an option to keep a small population of Delacour's langurs.

To ensure the survival of the species all previously suggested possibilities should be used:

1.   Improvement of protection for the largest and stable wild populations.

2.   Translocation of small populations and single groups without chance for long term existence to a well protected area with or without an existing population or to a semi-wild enclosure.

3.   Captive breeding with confiscated, hand raised or injured animals with the option to release these animals into a semi-wild enclosure.

3.5. Grey langur

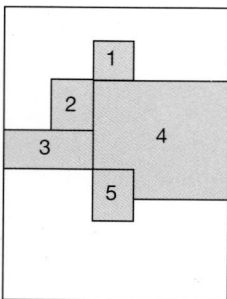

1. Grey langur; infant female, one year old. *T. Nadler*

2. Grey langur; ad. female with infant female three weeeks old. *T. Nadler*

3. Grey langur; photo trapping. *B. Long* (SFNC/FFI)

4. Grey langur; infant female one year old. *T. Nadler*

5. Grey langur; ad. male. *T. Nadler*

3.5   Grey langur
Trachypithecus crepusculus (Elliot, 1909)

3.5.1   Taxonomy

A grey langur from Mount Muleiyit, Burma was described as *Presbytis crepuscula* by Elliot (1909). Because of its similar colouration to the Phayre's langur, also described for Burma as *Presbytis phayrei* (Blyth, 1847), the taxa was placed subsequently as a subspecies to this langur in the revised genera systematic as *Semnopithecus phayrei crepuscula* (Corbet & Hill, 1992) or *Trachypithecus phayrei crepusculus* (Eudey, 1996/1997; Wang Yingxiang *et al.*, 1998; Groves, 2001).

Over the large area where Phayre's langurs occur, they show a wide variation in pelage colouration, and three (Groves, 2001) to five (Eudey, 1996/1997) subspecies have been recognized. But the relationship between the taxa belonging to the *obscurus/phayrei* group with very similar colouration was disputed (Osgood, 1932; Brandon-Jones, 1984; Lekagul & McNeely, 1988).

Recent DNA investigation (see 2.3; Roos *et al.*, 2001) cleared up the relationship of the Vietnamese taxon and identified this langur unambiguously as a distant relative of the *francoisi* group and not as a Phayre's langur subspecies.

Based on the Vietnamese common name for this species we propose to use the name Grey langur.

3.5.2   Description

The body and tail are light grey, the belly is slightly brighter and silvery grey. The forehead, cheeks and hair around the face are brown. Hands and feet are dark grey. The grey color on the hands changes along the forearms to the light grey of the body. The tail ends with a thin tassel. The head has no crest on the crown but long hairs in a whorl. The bare skin of the face is dark grey, except the light grey rings around the eyes and depigmented patches in the middle of the upper and lower lips.

External measurements (adult)

	n	mm	average	Source
Head/Body length				
male	1	510	510	EPRC
female	1	490	490	EPRC
Tail length				
male	1	830	830	EPRC
female	1	822	822	EPRC
Weight		kg		
male	1	6,9	6,9	EPRC
female	1	6,4	6,4	EPRC

3.5.3   Distribution

The Grey langur is distributed from central Thailand to north-west Thailand and south Yunnan, east to southwest Laos and northern Vietnam, and west to the coast of the Bay of Bengal, south of the range of *Trachypithecus p. phayrei* (Groves, 2001)

DISTRIBUTION OF GREY LANGUR (Trachypithecus crepusculus) IN VIETNAM
BẢN ĐỒ PHÂN BỐ LOÀI VOỌC XÁM Ở VIỆT NAM

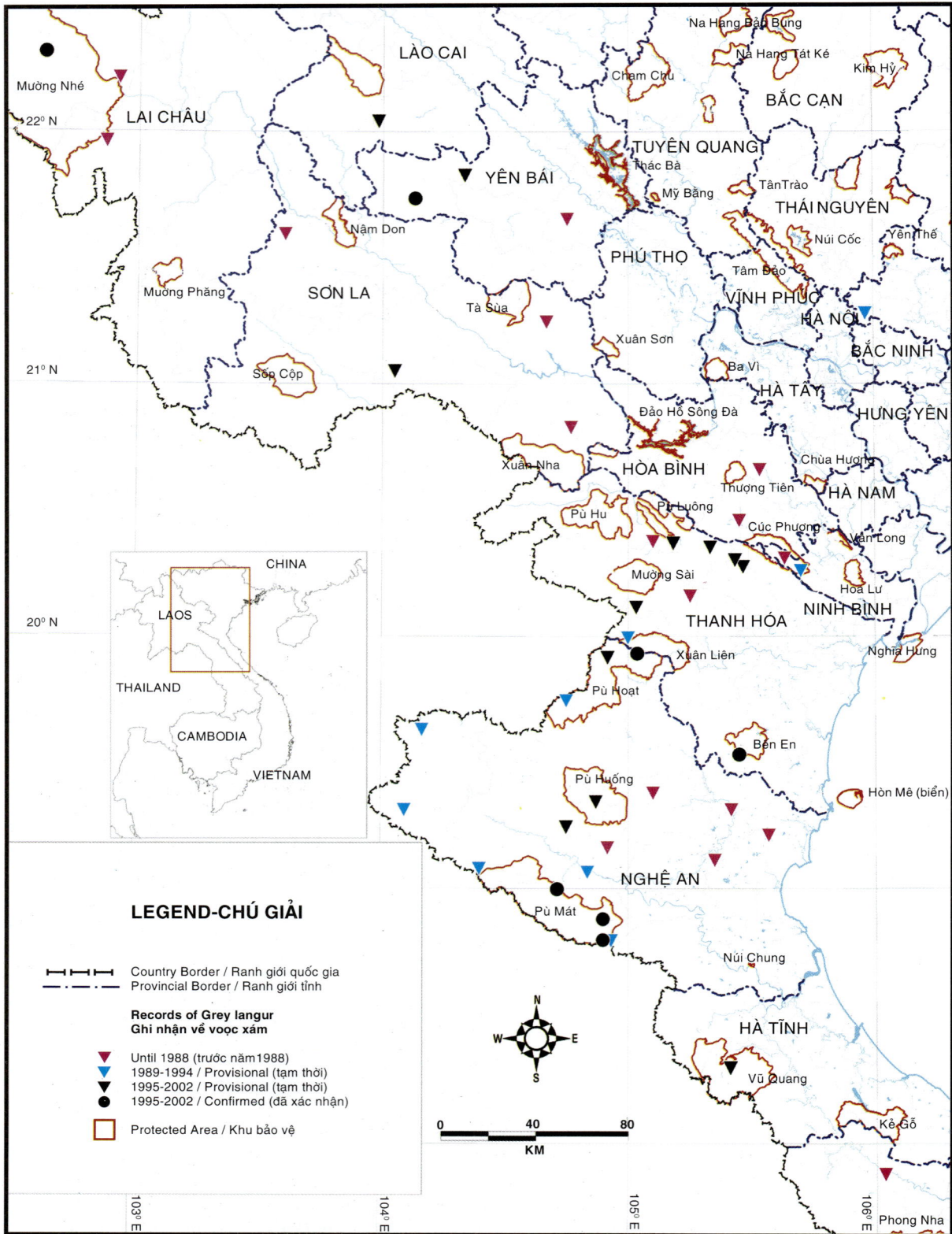

LEGEND-CHÚ GIẢI

Country Border / Ranh giới quốc gia
Provincial Border / Ranh giới tỉnh

Records of Grey langur
Ghi nhận về voọc xám

Until 1988 (trước năm 1988)
1989-1994 / Provisional (tạm thời)
1995-2002 / Provisional (tạm thời)
1995-2002 / Confirmed (đã xác nhận)

Protected Area / Khu bảo vệ

Distribution in China

T. crepusculus is restricted to south-east part of Yunnan (Xishuangbanna), which is the most northern part of its range (Wang Yingxiang *et al.*, 1998). It was recorded in Honghe County, Xishuangbanna, Baoshan and Yinjiag (Zhang *et al.*, 1992).

In the south-western part of Yunnan occurs the taxa *shanicus*. There is currently no detailed morphological and molecular genetic study of the systematic position of this taxa.

Distribution in Lao PDR

The range of *T. crepusculus* is widespread in the northern part of the country from the Mekong valley up to at least 800m a.s.l. Its occurrence is confirmed as far as Nam Kading NBCA and Phou Khaokhoay NBCA and provisionally as far south as Hin Namno NBCA (Duckworth *et al.*, 1999).

Distribution in Vietnam

The Grey langur inhabits the north-western part of Vietnam. The delineation of its range is unknown. Its occurrence is currently confirmed only west of the Black River (Song Da) and as far as Pu Mat National Park to the south. Its distribution seems to be highly fragmented. Other records are only based on interviews or collected specimens. It must be noted that data provided by local informants cannot be considered as adequate evidence of known presence. This species is often confused with macaque (*Macaca*) species by local people.

3.5.4 Grey langur records in Vietnam

Muong Nhe Nature Reserve (LAI CHAU)
Special use forest: Nature reserve
Grey langur status: Occurrence confirmed, last evidence in 1997 (Hill *et al.*, 1997)

One adult female (FMNH 31755) was collected in March 1929 by R. W. Hendee in Muong Mo Commune (east of the reserve). One male skin (IEBR 84) and one sub-adult male skull (IEBR 97/955) were collected in April 1963 by Liu Van Chem in Cha Cang Commune (south of the reserve) (Fooden, 1996).

Between January and March 1997, Frontier-Vietnam visited the southern part of Muong Nhe (Hill *et al.*, 1997). A single crippled individual was observed beside the Nam Ta Na (river), Muong Lay District.

Van Ban District (LAO CAI)
Special use forest: None
Grey langur status: Provisional occurrence, last report in 1999 (Dong Thanh Hai & Lormée, 1999)

FFI survey in Nam Xe and Nam Xay Communes in November 1999 reported, on the basis of interviews, the occurrence of Grey langur in Ho Nam Mu forest (Dong Thanh Hai & Lormée, 1999). Local informants suggested that only one group with 10 to 15 individuals lives in this area. This report was obtained again in December 2000 by a follow-up survey by FFI (Le Trong Dat *et al.*, 2000). Reports of this species were obtained by FFI teams from adjacent forest areas in 2001; one group, possibly of this species in Nam Xay Commune (B. Long, pers. comm) and one group in Khanh Yen Ha Commune (Ngo Van Tri, 2001).

Na Ot and Phieng Pan Communes, Mai Son District (SON LA)

Special use forest: None

Grey langur status: Provisional occurrence, last report in 1999 (L. Baker, pers. comm., 2000)

Grey langur was clearly described during a FZS survey for Delacour's langur in the area (L.Baker, pers. comm., 2000). In Na Ot, the last sighting was reported in January 1999. In Phieng Pan, nearly all forest has been cut down, and the species was reported to be still present but rare.

Thuan Chau, Phu Yen and Moc Chau Districts (SON LA)

Special use forest: None

Grey langur status: Unknown, last report in 1971 (Dao Van Tien, 1978)

The occurrence of Grey langurs in these districts is mentioned without detailed information. One male specimen was described by Dao Van Tien (1978), which was reportedly collected in July 1969 in Ban Tin Toc (collector and museum unknown).

Che Tao-Nam Pam Communes, Mu Cang Chai District (YEN BAI) and Muong La District (SON LA)

Special use forest: None

Grey langur status: Occurrence confirmed, last evidence in 2001 (Tordoff et al., 2001)

During a FZS survey for Delacour's langur, the occurrence of Grey langur in Nam Pam Commune was recorded on the basis of interviews (Baker, 1999b). The species is locally called "Ca la" or "Tu ca hang" and was recently sighted in groups of 20 (L. Baker, pers. comm. 2000). A survey by an FFI team in October 1999 did not collect any information about the species in this locality (Ngo Van Tri & Long, 1999). However, in the adjacent commune of Che Tao, Grey langurs were reported by local informants to live in the narrow bands of forest in the steep valleys (Nguyen Xuan Dang & Lormée, 1999; Long et al., 2000a).

The occurrence of the species was finally confirmed during a joint FFI-BirdLife survey of the locality in April 2001. Firstly, local people reported that 2 groups of Grey langur still occur near Na Hang village; one group of 7-8 animals in the Nang Lu valley and one group of 15-17 individuals in the Chua Lu valley. A group of four individuals was observed including one golden infant along the Nang Lu stream. Two adults were observed 15 minutes later and were possibly from the same group (Tordoff et al., 2001).

The local guide then reported observing a group of up to 30 langurs including 7 infants in the Chua Lu valley on the same afternoon at 2000m. a.s.l.. Given that local people reported that the larger group consisted of only 15-17 individuals, this may be an exaggeration. However, the two observations confirm the presence of the species in this locality, and, furthermore, it is almost certain that these are separate groups. Throughout the survey work in the Che Tao forests five groups of T. crepusculus have been confirmed or reported in totalling a minimum of 29 animals, but probably many more (Long et al., 2001). This is a significant sub-population of this species, however, hunting may be a considerable threat. In 1993, one hunter apparently shot 9 animals on one occasion. If this event is repeated a small number of times, the species may become extinct in this locality (Tordoff et al., 2001).

Cau Pha forest, Mu Cang Chai and Van Chan Districts (YEN BAI)
Special use forest: None
Grey langur status: Provisional occurrence, last report in 1999 (Nguyen Xuan Dang & Lormée, 1999)

In October 1999, interviews were conducted during a short survey by an FFI team to assess the primate status in the area (Nguyen Xuan Dang & Lormée, 1999). Grey langurs were reported by local informants.

Phong Du Thuong, Lam Giang and Xuan Tam Communes, Van Yen District (YEN BAI)
Special use forest: None
Grey langur status: Provisional occurrence, last report in 1999 (Nguyen Xuan Dang & Lormée, 1999)

In October 1999, interviews were carried out by a FFI team in the district town of Van Yen (Nguyen Xuan Dang & Lormée, 1999). Local informants, who often go into the forest of Phon Du Thuong and Xuan Tam Communes, reported the occurrence of Grey langurs. However, due to the lack of time, these places were not visited by the team and the information could not be confirmed.

Lam Giang Commune was visited by an FFI team in October 1999 to assess the status of primates in the area (Nguyen Xuan Dang & Lormée, 1999). Grey langurs were reported by local informants to have been seen occasionally in groups of 10 to 15 individuals in Nui Con Voi (mountain) forest.

Tran Yen District (YEN BAI)
Special use forest: None
Grey langur status: Unknown, last report before 1994 (Dang Huy Huynh *et al.*, 1994)

The occurrence without details is only mentioned by Dang Huy Huynh *et al.*, (1994).

Ban Tin Toc, Moc Chau District (SON LA)
Special use forest: None
Grey langur status: Unknown, last report in 1969 (museum specimen)

One male specimen was described by Dao Van Tien (1978), which was reportedly collected in July 1969 (N°357, collector and museum unknown) in Ban Tin Toc.

Kim Boi District (HOA BINH)
Special use forest: Partly included in Thuong Tien Nature Reserve
Grey langur status: Unknown, last report in 1989 (museum specimen)

One male skin (FCXM unnumbered) was purchased in 1989 (collector unknown) in Kim Boi District (Fooden, 1996).

Ngoc Son Area, Lac Son District (HOA BINH)
Special use forest: None
Grey langur status: Unknown, last report in 1973 (museum specimen)

One sub-adult male skull (IEBR 118) was collected in April 1973 by Pham Trong Anh and reported from Ngoc Lan locality (Fooden, 1996).

Pu Luong Nature Reserve (THANH HOA)
Special use forest: Nature Reserve
Grey langur status: Provisional occurrence, last report in 1999 (Baker, 1999b)

Two adult males (BMNH 1933.4.1.7 and 1933.4.1.8) were collected in January-February 1930 by J. Delacour and W. P. Lowe in Hoi Xuan Commune (Napier, 1985). One juvenile female (IEBR 679/530(12)) and one female skin (IEBR 683/534(16)) were collected in March 1964 by Ha Van Thau in the same locality (Fooden, 1996).

During a FZS survey conducted from February to June 1999, with a special emphasis on the Delacour's langur, information provided by local informants in different places (Lung Cao, Co Lung, Thanh Son, Hoi Xuan, Thanh Xuan and Phu Le Communes) reported the occurrence of Grey langur in Pu Luong Nature Reserve (Baker, 1999b).

Nui Tuong limestone area, Ba Thuoc District (THANH HOA)
Special use forest: None
Grey langur status: Provisional occurrence, last report in 1999 (Baker, 1999b)

A FZS survey conducted from February to June 1999 reported the occurrence of *T. crepusculus*, locally named "zan", in Luong Noi Commune, based on interviews (Baker, 1999b).

Local informants reported the species as extremely rare during the survey conducted by FFI in the area in October 1999 (Ngo Van Tri, 1999a). A hunter from Am Village claimed to have seen a troop of seven animals on the top of a big tree in Cam Valley in 1997.

Cam Quy Commune, Cam Thuy District (THANH HOA)
Special use forest: None
Grey langur status: Provisional occurrence, last report in 1999 (Baker, 1999b)

Grey langurs, locally named "zan" are reported, based on interviews during a FZS survey conducted from February to June 1999 (Baker, 1999b)

Thach Tuong Commune, Thach Thanh District (THANH HOA)
Special use forest: None
Grey langur status: Provisional occurrence, last report in 1999 (Baker, 1999b)

Grey langurs, locally named "zan" are reported, based on interviews during a FZS survey conducted from February to June 1999 (Baker, 1999b). The forest extends to the westernmost tip of Cuc Phuong National Park.

Lang Chanh District (THANH HOA)
Special use forest: None
Grey langur status: Unknown, last report in 1964 (museum specimen)

One adult female (IEBR 634/486(33A)) and one juvenile female (IEBR 635/487(34A)) were collected in March 1964 in the Tan Phuc locality (probably by Lo Van Let). One adult male (IEBR 625) was collected in March 1964 by Le Van Soat and reported from Lang Chanh. One adult skull, possibly female (IEBR 646), was collected from the same locality in March 1964 (unknown collector) (Fooden, 1996).

Xuan Lien Nature Reserve (THANH HOA) / Pu Huot Nature Reserve (NGHE AN)
Special use forest: Nature reserve
Grey langur status: Occurrence confirmed, last evidence in 2002 (La Quang Trung & Trinh Dinh Hoang, 2002)

The occurrence of Grey langurs was reported in Thuong Xuan District in 1989 by Ratajszczak *et al.* (1990) based on interviews. The occurrence was also provisionally reported during the feasibility study conducted in October and November 1998 by BirdLife and FIPI (Le Trong Trai, 1999a) to establish Xuan Lien Nature Reserve in the western part of Thuong Xuan District. Body parts of Grey langur were seen in hunter's house by a FIPI team in Pu Hoat NR, in July 1997 (Le Trong Trai, pers. comm. 2000). La Quang Trung & Trinh Dinh Hoat (2002) confirmed by sighting of 6 individuals the presence of Grey langur on the border between Xuan Lien NR and Pu Huot NR in April 2002.

Ben En National Park (THANH HOA)
Special use forest: National park
Grey langur status: Occurrence confirmed, last evidence in 1997 (Tordoff *et al.*, 1997)

During a survey between July and September 1997, a SEE team observed two individuals drinking at a stream in the south of the park (Tordoff *et al.*, 1997).

Nghia Dan District (NGHE AN)
Special use forest: None
Grey langur status: Unknown, last report in 1928 (museum specimen)

Two adult males (BMNH 1928.7. 1.6 and MNHN 1929. 436) were collected in February 1928 by J. Delacour and W. P. Lowe and were reported from the Nghia Dan locality (Fooden, 1996; J. M. Pons, pers. comm. 2000).

Nghia Dung, Tan Ky District (NGHE AN)
Special use forest: None
Grey langur status: Unknown, last report in 1964 (museum specimen)

One adult male (IEBR 712), one adult female (IEBR 713) and one infant (IEBR 716) were collected in December 1964 by Lo Van Chuong in the Nghia Dung locality (Fooden, 1996).

Que Phong District (NGHE AN)
Special use forest: None
Grey langur status: Unknown, last report in 1989 (Ratajszczak *et al.*, 1990)

According to interviews conducted during a primate survey in 1989 (Ratajszczak *et al.*, 1990), Grey langurs may occur in the western part of the district.

Ky Son District (NGHE AN)
Special use forest: None
Grey langur status: Unknown, last report in 1989 (Ratajszczak *et al.*, 1990)

Very little forest remains in the district. The area north of Song Ca (river) has been nearly totally deforested. The only good forest apparently remains in the southern part of the district, in the vicinity of Phu Xai Lai Leng Mountain, along the Lao PDR border (Rozenddal, 1990).

The locality is listed in Ratajszczak *et al.* (1990). Interviews conducted in 1989 report the occurrence of Grey langurs close to the Lao border.

Pu Huong Nature Reserve (NGHE AN)
Special use forest: Nature reserve
Grey langur status: Provisional occurrence, last report in 1995 (Kemp & Dilger, 1996)

One male was collected in November 1974 (collector and museum unknown) in Ke Can locality, Quy Chau District, east to Pu Huong (Dao Van Tien, 1985).

Grey langur was listed as present in Pu Huong (=Bu Huong) in primary and secondary forest by a SEE survey (Kemp & Dilger, 1996). This report was based on interviews conducted between April and June 1995.

Pu Mat National Park (NGHE AN)
Special use forest: National park
Grey langur status: Occurrence confirmed, last evidence in 1999 (SFNC/FFI, 2000)

Ratajszczak *et al.* (1990) reported, on the basis of interviews, that Grey langurs might occur in Tuong Duong and Anh Son Districts, close to the Lao border. A SEE team, which carried out a survey in Con Cuong district in November-December 1994 (Kemp *et al.*, 1995b) reported the species from interviews. Grey langurs were said to live in both primary and secondary forest. During surveys conducted for the SFNC project the species presence was confirmed. Six animals were seen on the lower slopes in secondary forest, west of Huoi Chat (central north of the reserve) in July 1998. In October and November 1998, two animals and then one single individual were phototrapped in Khe Khang sector (southeast of the reserve). In April 1999, one troop of 5 individuals, presumed to be this species, was observed in the Cao Veu area (southernmost part of the reserve) and another sighting of a group of at least 4 animals in the Khe Bong valley in July 1999 (SFNC/FFI, 2000).

The absence or extremely low densities of Grey langur in many suitable habitats indicates an extremely high and selective hunting pressure (Johns, 1999). This is demonstrated by the estimate of 900 people per day entering the core zone illegally (SFNC/FFI, 2000).

The EPRC received two Grey langurs confiscated in Pu Mat in April and November 2000 (Nadler, pers. comm.).

Quynh Chau Commune, Quynh Luu District (NGHE AN)
Special use forest: None
Grey langur status: Unknown, last report in 1962 (museum specimen)

One adult female (ZMVNU 108/Ps/30) was collected in this locality in October 1962 by Vu Thanh Tinh (Fooden, 1996).

Vu Quang National Park (HA TINH)
Special use forest: National park
Grey langur status: Provisional occurrence, last report in 1997 (Anon., 1997b)

In 1995 and July-August 1997, two surveys were carried out by a VRTC expedition for the Vu Quang Conservation Project (WWF) in the reserve. Grey langurs were reported based on interviews (VRTC, 1997).

Tuyen Hoa District (QUANG BINH)
Special use forest: None
Grey langur status: Unknown, last report unknown (museum specimen)(Fooden, 1996)

One male skull (IEBR 44) is reported to have been collected in Tuyen Hoa District (date and collector unknown) (Fooden, 1996). The occurrence of the Grey langur was not reported by Lambert *et al.* (1994) during their survey for endemic pheasants in the Khe Net area (north of the district).

Locations where Grey langurs are believed to be extinct

Phu Yen District (SON LA)

One adult female (IEBR 843/221(44)) and one male skin (IEBR 844/222) were collected in Phieng Ban locality by Vi Van Nao in October 1963 (Fooden, 1996). In October-November 1999, a FFI team visited the district (Ngo Van Tri & Long, 1999). A number of interviews were conducted (fifteen) but no information was provided concerning the Grey langur.

Cuc Phuong National Park (NINH BINH, HOA BINH and THANH HOA)

Two skins, one adult female (ZMVNU 82) and one infant female (ZMVNU 83) were collected in October 1960 (collector unknown) in Cuc Phuong area (Fooden, 1996). Ratajszczak *et al.*, (1990) reported about two captive Grey langurs in Cuc Phuong National Park in 1989. These animals were captured from a group of four individuals, while they were drinking in a stream on agriculture land, 500 meters from "Phu Long Village", Nho Quan (=Hoang Long) District (according to the map provided in the report, located in the border southeast of Cuc Phuong National Park). The local informants said that no animals had been seen for many years prior to this.

Cuc Phuong has been the subject of a number of surveys. Despite this, there was no further records of Grey langur in the park (Nadler, 1995a; Hill, 1999; T. Nadler, pers. comm.).

Pieng Luong village, Phong Lai Commune, Thuan Chau District (SON LA)

The species has not been seen for 15 years (interviews conducted during a FZS survey in 1999, L. Baker, pers. comm., 2000).

Chieng Set village, Chieng Den Commune, Son La District (SON LA)

The species was last seen nearly 30 years ago (interviews conducted during a FZS survey in 1999, L. Baker, pers. comm., 2000).

Ban Dan and Ban Phang villages, Muong Bang Commune, Mai Son District (SON LA)

Grey langur was described but had not been seen for 6 years (interviews conducted during a FZS survey in 1999, L. Baker, pers. comm).

Ban Puon village, Chieng Mai Commune, Mai Son District (SON LA)

The last sighting was reported several years ago, and the species is probably now extinct in the area (interviews conducted during a FZS survey in 1999, L. Baker, pers. comm., 2000,).

Ban Tam and Tu Buon villages, Chieng Ve Commune, Mai Son District (SON LA)

The species is described but last sightings are reported 10 to 20 years ago (interviews conducted during a FZS survey in 1999, L. Baker, pers. comm., 2000).

Discussed records

Ky Thuong Nature Reserve (QUANG NINH) 21°05'-21°12'N/106°56'-107°13'E

Grey langurs are reported in the investment plan of Ky Thuong Nature Reserve, Quang Ninh Province. A record so far to the east is questionable. During a visit to the reserve by BirdLife and FIPI in November 1999, seven informants claimed that no langurs occurred in the area (Tordoff *et al.*, 2000b). However, a Dao hunter reported to have seen this species ("Vooc xam") in 1959, though it is now extinct.

Na Hang Nature Reserve (TUYEN QUANG) 22°16'-22°31'N/105°22'-105°29'E

Grey langur is reported based on interviews by Dang Huy Huynh & Hoang Minh Khien (1993) and Boonratana (1999). This occurrence can be questioned. According to B. Martin (pers. comm. 2000), information provided by local people is confused and does not provide good evidence and it is likely that, if the species ever existed in Na Hang, it does not occur anymore. However, some areas in Na Hang are quite inaccessible and the fauna living in these places is still unknown.

Du Gia Nature Reserve (HA GIANG) 22°49'-22°58'N/105°03'-105-11'E

Grey langurs were reported in the investment plan to establish Du Gia Nature Reserve (Anon., 1994c). However, no references are given. Therefore this record is considered unreliable.

Phong Quang Nature Reserve (HA GIANG) 22°50'-23°04'N/104°50'-105-01'E

Grey langurs are reported in the investment plan for Phong Quang Nature Reserve (Anon., 1997b). However, no reference is given.

Pa Co-Hang Kia Nature Reserve (HOA BINH) 20°41'-20°46'N/104°51'-105°01E

Grey langurs are reported in the feasibility study for Pa Co-Hang Kia Nature Reserve (Anon., 1993c). However, no reference is given. Therefore, this record cannot be considered as reliable.

Ha Tay Province 20°34'-21°17'N / 105°18'-105°58'E

Grey langurs were reported in Ha Tay Province in the *Red Data Book of Vietnam* (Ministry of Science, Technology and Environment, 1992). However, this location is given without reference. Regarding the extent of forest currently remaining in the province, it is doubtful that a significant population of the species still occurs.

Thanh Son District (PHU THO) ca. 21°21'N / 104°59'E

Grey langurs were reported in Thanh Son District (Phu Tho Province) by Dang Huy Huynh *et al.* (1994). However, no reference is given and the information cannot be considered as reliable. The species was not recorded during the FFI survey in Xuan Son Nature Reserve in November 1999 (Nguyen Xuan Dang & Lormée, 1999).

Kon Cha Rang - Kon Ka Kinh area (GIA LAI) 14°09'-14°35'N/108°16'-108°39E

The report of Grey langur in Kon Cha Rang-Kon Ka Kinh area by Lippold (1995b) is considered to be "incongruous" by Fooden (1996). This occurrence would be a considerable extension of *T. crepusculus* range to the south and there must be repeated sightings before it can be classified as confirmed.

3.5.5 Status

Not enough data are available to assess correctly the Grey langur status in Vietnam. In fact, no surveys, however short, have been carried out with a special emphasis on this species.

Grey langur is now very scarce and shy in Vietnam. In ten years, only five confirmed records have been documented, and there are only 12 localities with provisional occurrence.

In China, the range of the species is restricted. In Honghe County it is extirpated from most places (Zhang *et al.*, 1992). Ma *et al.* (1988) gave an estimation of 11,500 to 17,000 individuals. If this number was not overestimated, the population in China has dropped to 5,000-6,000 individuals (Wang Yingxiang *et al.* 1998). Zhang Yongzuo *et al.* (2002) estimate 5,400 individuals (of both taxa, *shanicus* and *crepusculus*). The subspecies *shanicus* occurs in three protected areas, the subspecies *crepusculus* in eight protected areas.

No research on the species is planned at the moment in the country (Li Zhaoyuan, pers. comm., 2000).

Despite its wide distribution in Lao PDR, there are less confirmed records of the Grey langur than of any other primate species. Although Nam Et, Phou Louey and Nam Ha NBCAs are located in its range, it has not been recorded, even provisionally. Its occurrence was recently confirmed in only three locations. Furthermore, it has not been found common in any surveyed area. It may be the most threatened primate in Lao PDR. It is locally considered **"At Risk"** (Duckworth *et al.*, 1999).

The major threat it currently faces is from hunting. Grey langur is heavily hunted throughout their range. This species often develops gallstones as large as golf balls, possibly due to drinking in the salt springs of mountainous areas. These "bezoar" stones are much sought-after by the Chinese for medicine production (Lekagul & McNeely, 1988).

Indeed, although the species can live in a disturbed habitat, where it may even be found in higher densities (MacKinnon & MacKinnon, 1987), the hunting pressure is so great in northern Vietnam that its population can be considered to be on the way to extinction. Being a large prey, it is a prime target for hunters and, like other primates, it constitutes valuable merchandise for wildlife trading.

The *IUCN Red List of Threatened Species* (Hilton-Taylor, 2000) does include the species.

The Phayre's langur was listed by Eudey (1996/1997) with the IUCN Red List category **"Data Deficient"**

The species (named as *Trachypithecus phayrei crepusculus*) is currently considered as **"Vulnerable"** in Vietnam (Pham Nhat *et al.*, 1998). However, though widely distributed, the species is now present in very low densities and in few isolated localities. The most recent data suggests that consideration of the species as **"Critically Endangered" CR** A1cd, A2cd, C2a best reflects the reality of the situation in Vietnam.

3.5.6   Recommendations for conservation in Vietnam

As for other primate species, general recommendations, such as reinforcement of hunting and wildlife trade control must be undertaken by Vietnamese authorities to ensure the survival of Grey langurs in Vietnam.

However, since we have no detailed information about the status and the distribution of Grey langur in Vietnam, few recommendations focused on one location can be suggested.

3.5.6.1.   Conduct further field status surveys

Further surveys are needed to assess the status of this species, which is poorly known about. In fact, only a few areas in Vietnam are known to contain any *T. crepusculus*.

Although no data is available from this area, Ky Son District (Nghe An Province) may be a good target for conservation of this species. Indeed, according to the forest cover maps, this area may possess the largest expanse of forest in its range in Vietnam.

Furthermore, Pu Mat National Park, where Grey langurs were recently observed, is adjacent to this region. It requires primate focused surveys.

There are still a number of areas of the Pu Mat National Park where no data on primate densities have been collected and which might be reasonably undisturbed. These include the area south of the Khe Bu River and the northern sector of the park, north and northwest of the Khe Thoi River. It is recommended that additional surveys be conducted in these areas to determine the status of the Grey langur population (Johns, 1999).

3.5.6.2.   Reinforcement of conservation in existing protected area network

The species is known to occur in Muong Nhe Nature Reserve. However, as it stands, the reserve is unable to support any substantial protection of the species. Although it is the largest protected area in Vietnam, only 20% of the land is actually covered by forest, which is highly fragmented. The boundaries should be reviewed in order to include the remaining patches of natural habitat and to exclude most of the grassland, settlements and fields. In addition, according to Duckworth *et al.* (1999), Phou Dendin NBCA, which is located in the adjacent region in Lao PDR, is one of the few places where this species may be locally numerous.

The low density of Grey langurs, and primates in general, in Pu Mat National Park is alarming. In the central areas, the forest is still more or less intact due to inaccessibility. Here, primate densities are very low because of extremely high hunting pressure. Approximately one hunting camp per km of main river inside the reserve has been estimated by researchers working in the reserve. It is recommended that the existing regulations for the protection of the reserve be strictly enforced. Most urgent are the confiscation of weapons and the arrest of professional hunters (Johns, 1999).

Further survey should be undertaken in Xuan Lien Nature Reserve and adjacent Pu Huot Nature Reserve to develop primate species specific conservation interventions.

3.6. Indochinese Silvered langur

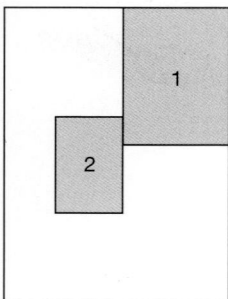

1. Indochinese Silvered langur; ad. male. *T. Nadler*
2. Indochinese Silvered langur; ad. female. *T. Nadler*

3.6   Indochinese langur
Trachypithecus germaini (Milne-Edwards, 1876)

3.6.1   Taxonomy

The taxonomy of the Silvered langur has been under debate for a long time. Milne-Edwards (1876) described a Silvered langur for Cochin China and Cambodia as *Semnopithecus germaini*. Due to similar colouration the taxa was subsequently placed by many authors as a subspecies to the Silvered langur from Sumatra, described as *Simia cristata* by Raffles (1821) and later assigned to several genera *Presbytis*, *Semnopithecus* and *Trachypithecus*.

Ellerman & Morrison-Scott (1951) recognize *T. c. germaini* as the sub-species occupying southern Vietnam, Lao PDR and Cambodia. This nomenclature is adopted by P. Napier (1985) who argues that specimens of *T. c. germaini* are generally lighter in colour, and have a different colour distribution of hair and whiskers than the typical *T. c. cristatus*.

Lekagul & McNeely (1988) mention the different colouration of Silvered langurs in Thailand and assumed three subspecies for the country.

Elliot (1909) described *Presbytis margarita* from the Long Bian Plateau (12°03'N) of Lam Dong Province, though neither Osgood (1932), Pocock (1935) and Groves (2001) found it sufficiently different from *germaini* to warrant sub-specific distinction.

Dao Van Tien (1977) described the sub-species *T. c. caudalis*, based on two animals of unknown origin which had been living in Hanoi zoo. Later (undated MS) he localised the origin to Tuyen Hoa District, Quang Binh Province (17°53'N) on the basis of pelage colouration. The type and paratype are preserved in the ZMHNU. Groves (2001) accepts this subspecies and notes the similarity of these specimens to one from Thailand in the USNM collection.

Dao Van Tien (undated MS) suggests a subspecific splitting of the species based on colour variation: *T. c. germaini* in the southern lowlands, *T. c. margarita* on the Lang Bian Plateau and *T. c. caudalis* in the central highlands up to Quang Binh. But he did not provide sufficient evidence for this splitting.

Groves (2001) assigns all the Indochinese Silvered langurs to *T. germaini*, reserving *T. cristatus* for Sundaic forms on Borneo, Sumatra and along the west coast of the malaysian peninsula. He pointed out that, beside different colouration and circumfacial hair structure, the length of the tail between these two groups is markedly different.

Recent DNA analysis supports this classification and shows the diffentiation at species level (see 2.3).

However, animals recently sighted in Dak Lak Province (Ngo Van Tri, 2000), showed a particularly dark colouration and Hoang Minh Duc (pers. comm., 2000) found populations with different colouration in south Vietnam. Elliot (1912) mentioned already the different colouration of two animals in MNHN Paris, both are marked as types. Despite of no marked DNA differences, the Vietnamese population shows a wide variety in colouration or probably a polymorphism of the taxon. The related taxa *cristatus* and *auratus* also have rather different morphs.

We propose to use the name Indochinese Silvered langur for *Trachypithecus germaini*.

3.6.2   Description

The colour is medium grey, caused by short creamy tips on the dark grey or brown-black hairs. Underside, throat, and shanks creamy, grading into upper side; hands and feet black, hairs form a creamy "bowl" around the face; tail nearly black above, lighter below.

External measurements (adult)

There is only scarce information about the measurement because of the confusion of the identification.

	n	mm	Average	Source
Head/Body length				
male	1	587	587	Flower, in Elliot, 1912
Tail length				
male	1	800	800	Flower, in Elliot, 1912
male/female	6	720 - 838	?	Groves, 2001

3.6.3   Distribution

The Indochinese Silvered langur occurs from Kanchanaburi west to the sea of Burma in a narrow strip; southeastern Thailand as far north as Loei; Cambodia; and southern Vietnam (Groves, 2001).

Distribution in Lao PDR

The species is widespread in southern Lao PDR, where it has been found in at least seven survey areas (Duckworth *et al.*, 1999). Its occurrence is confirmed north to Xe Bang-Nouan NBCA (Saravanne Province) at 15°45'N. Local people reported "grey langur" sightings from two sites north of the confirmed range. Indochinese Silvered langurs provisionally identified in the Nam Phoun NBCA (Boonratana, 1998) would represent a large northward extension of the known range and it's a mistake (Brandon-Jones in lit., 1999).

Distribution in Cambodia

The range of Indochinese Silvered langurs in Cambodia may be widespread (Corbet & Hill, 1992). During the last FFI surveys in the country (Timmins & Men Soriyun, 1998; Long *et al.*, 2000b; Long *et al.*, 2000c), the species was recorded in southern Ratanakiri Province, the Cardamom Mountains and in northeastern Mondulkiri Province. Distribution appeared to be patchy in the Cardamom Mountains and Silvered langurs were only found in lowland evergreen forest. In southern Ratanakiri and northeastern Mondulkiri Province, a large area dominated by a dry deciduous forest, distribution is fragmented and restricted to patches of riverine evergreen forest.

Distribution in Vietnam

T. germaini is known from Quang Tri Province at 16°37'N latitude to the Mekong delta. The occurrence of this species further north in Tuyen Hoa District (Quang Binh Province)(Dao Van Tien, undated MS) is disputed by Fooden (1996). Osgood (1932) found as questionable the record in Lao Bao (Quang Tri Province). In fact, there are no reliable reports in Vietnam between 16°37'N and 14°30'N.

Despite a number of surveys, the species has never been recorded in Bach Ma National Park (Thua Thien-Hue Province) and according to Do Tuoc (pers. comm., 2000), though structured interviews were carefully conducted in Quang Nam Province, no evidence of Indochinese Silvered langurs was reported.

DISTRIBUTION OF INDOCHINESE SILVERED LANGUR *(Trachypithecus germaini)* IN VIETNAM
BẢN ĐỒ PHÂN BỐ LOÀI VOỌC BẠC Ở VIỆT NAM

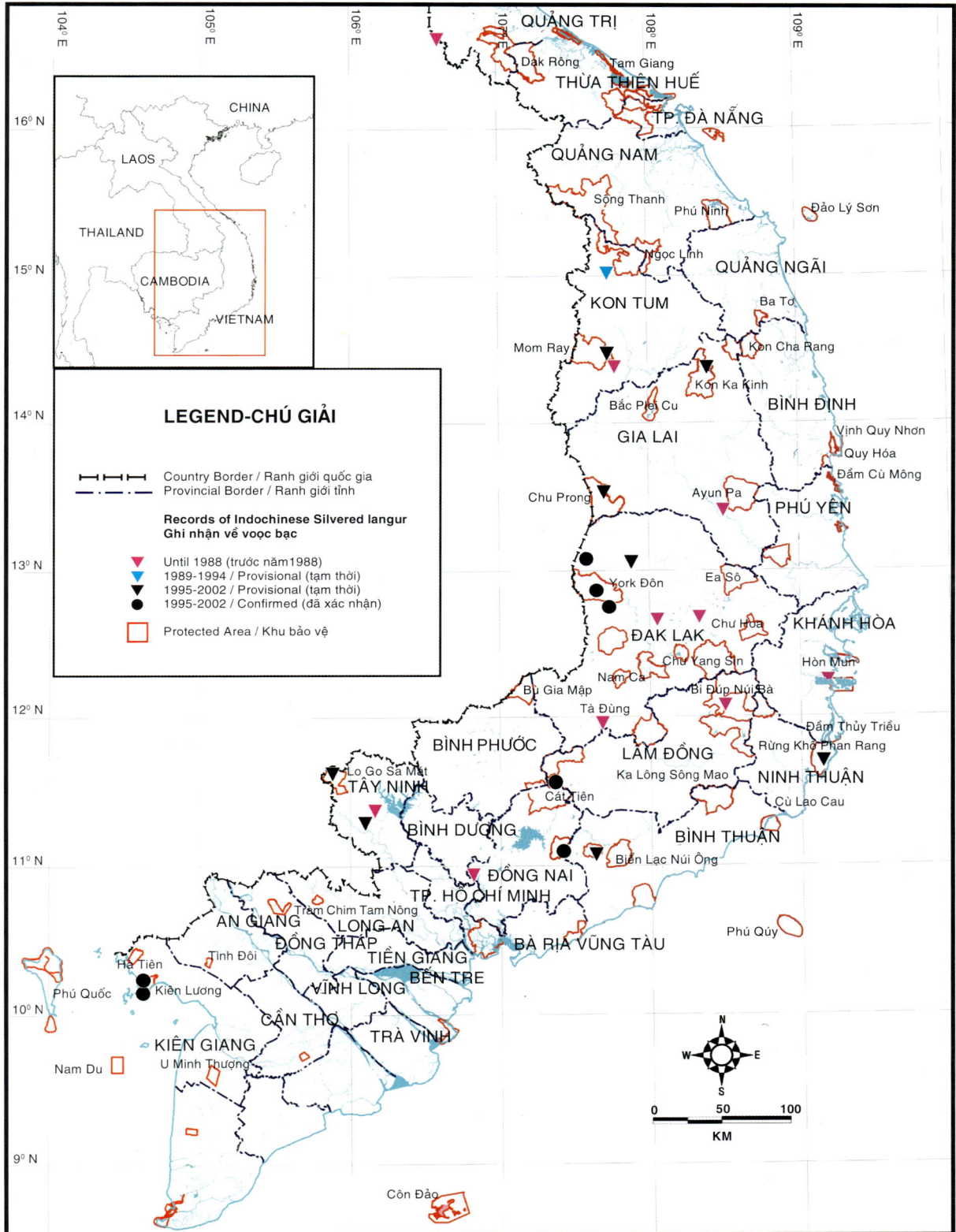

LEGEND-CHÚ GIẢI

Country Border / Ranh giới quốc gia
Provincial Border / Ranh giới tỉnh

**Records of Indochinese Silvered langur
Ghi nhận về voọc bạc**

Until 1988 (trước năm 1988)
1989-1994 / Provisional (tạm thời)
1995-2002 / Provisional (tạm thời)
1995-2002 / Confirmed (đã xác nhận)

Protected Area / Khu bảo vệ

3.6.4   Indochinese Silvered langur records in Vietnam

Lao Bao (QUANG TRI)
Special use forest: None
Indochinese Silvered langur status: Unknown, last report in 1929 (museum specimen) (Napier, 1985; Fooden, 1996)

One adult female skin (BMNH 1933.4.1.12), one juvenile male skin (BMNH 1933.4.1.13) and two male skins (FMNH 39157 and 39158) were collected in January 1929 by E. Poilane (Napier, 1985; Fooden, 1996). The reliability of this record was questioned by Osgood (1932).

Dac Glei District (KON TUM)
Special use forest: None
Indochinese Silvered langur status: Provisional occurence, last report in 1991 (museum specimen) (Fooden, 1996)

One female skin was collected in 1991 by Truong Son (FCXM unnumbered) (Fooden, 1996).

Mom Ray National Park (KON TUM)
Special use forest: National park
Indochinese Silvered langur status: Provisional occurence, last report before 1995 (Lippold, 1995)

The occurrence is mentioned without detailed information.

Sa Thay and Dac To Districts (KON TUM)
Special use forest: None
Indochinese Silvered langur status: Provisional occurrence, last report before 1994 (Dang Huy Huynh *et al.,* 1994)

The occurrence is mentioned without detailed information.

Chu Prong District (GIA LAI)
Special use forest: Partly included in Chu Prong Nature Reserve
Indochinese Silvered langur status: Provisional occurrence, last report in 2000 (specimen)(Tran Hieu Minh *et al.,* in prep.)

Remains of Indochinese Silvered langurs that had been hunted were seen by a BirdLife and FIPI team in April 2000. The area is dominated by a dipterocarp forest habitat.

Kon Ha Nung area (GIA LAI)
Special use forest: Partly included in Kong Cha Rang and Kon Ka Kinh Nature Reserves
Indochinese Silvered langur status: Provisional occurrence, last report in 1999 (specimen)(Le Trong Trai, pers. comm., 2000)

A feasibility study was conducted by FIPI to establish Kong Cha Rang Nature Reserve in April 1999 (Le Trong Trai, 1999b). One skull was seen by the team in a hunters' house. The animal was reported to have been shot in Tram Lap Forest Enterprise, between Kong Cha Rang and Kon Ka Kinh Nature Reserves (Le Trong Trai, pers. comm., 2000).

Cu M'lan, Ea Sup District (DAK LAK)
Special use forest: None
Indochinese Silvered langur status: Occurrence confirmed, last evidence in 1997 (Le Xuan Canh *et al.*, 1997a)

One group of about eight Indochinese Silvered langurs was seen near the Dak Rue River in June 1997 during a survey conducted by IUCN and WWF for large mammals (Le Xuan Canh *et al.*, 1997a). The species was not recorded by BirdLife and IEBR during a survey for the green peafowl in the area from March to May 1998 (Brickle *et al.*, 1998).

Yok Don National Park (DAK LAK)
Special use forest: National park
Indochinese Silvered langur status: Occurrence confirmed, last evidence in 1997 (Le Xuan Canh *et al.*, 1997a)

Indochinese Silvered langurs are reported, based on indirect evidence (no further details are given), during an elephant survey carried out by WWF in February-March 1992 (Dawson *et al.*, 1993). One group of at least four individuals was seen on the summit of Yok Don Hill in May 1997 during a survey conducted by IUCN and WWF for large mammals (Le Xuan Canh *et al.*, 1997a). This locality is not reported as holding Indochinese Silvered langur in a survey for green peafowl by BirdLife and IEBR in March to May 1998 (Brickle *et al.*, 1998).

Cu Jut District (DAK LAK)
Special use forest: partly included in proposed extension of Yok Don National Park
Indochinese Silvered langur status: Occurrence confirmed, last evidence in 1999 (Ngo Van Tri & Trinh Viet Cuong, 2000)

FFI conducted two elephant field surveys in the district. In December 1999, one group of at least 3 animals were seen in secondary evergreen forest in the western part of the district (12°44'47.7"N / 107°44'35.5"E) between 300 and 400 m a.s.l. (Trinh Viet Cuong & Ngo Van Tri, 1999). In April 2000, one group of 5 animals, including one infant with an orange/ yellow coat, was seen in semi-evergreen forest near the first record (12°45'09" N / 107°44'42"E) (Ngo Van Tri, 2000).

Da Lat Plateau (DAK LAK, LAM DONG, KANH HOA and NINH THUAN)
Special use forest: Partly included in Chu Yang Sinh and Bi Dup-Nui Ba Nature Reserves
Indochinese Silvered langur status: Unknown, last report in 1908 (museum specimen)

One juvenile male (BMNH 1908.11.1.5) was collected in 1908 by J. J. Vassal on Nui Ba Mountain (=Lang Bian) (Napier, 1985).

Cat Tien National Park (DONG NAI, BINH PHUOC and LAM DONG)
Special use forest: National park
Indochinese Silvered langur status: Occurrence confirmed, last evidence in 2002 (G. Polet, pers.comm.)

One reliable sighting of one group was reported on 15 June 2000 (J. Barr). Another group of 8 animals was seen on 19 April 2001 (D. Hobcroft). Another group of more than 10 animals including 2 babies was observed on 9 June 2002 (G. Polet, pers. comm.).

Sightings of the species without details are also noted by Pham Nhat *et al.* (2001)

Nui Chua Nature Reserve (NINH THUAN)
Special use forest: Nature reserve
Indochinese Silvered langur status: Provisional occurrence, last report in 2000 (specimen)(Ha Thang Long, 2001)

During a SIERES survey of the locality in April 2000, one animal was found in the house of a wildlife trader near to the headquarters of the nature reserve (Vu Ngoc Long, 2001). However, this is the only information from this area. The species was also not reported during a FZS survey in August 2001. It is unlikely that the species occurs in the nature reserve.

Bien Lac-Nui Ong Nature Reserve (BINH THUAN)
Special use forest: Nature reserve
Indochinese Silvered langur status: Provisional occurrence, last report in 1998 (specimen) (Ngo Van Tri, 1999b)

FFI conducted a short field survey in September 1999 to assess the status of elephants in the nature reserve. The team saw one Indochinese Silvered langur trophy shot from a group of 7 to 11 animals by a policeman in April 1998. However, the occurrence could not be confirmed. (Ngo Van Tri, 1999b)

Dinh Quan District (DONG NAI)
Special use forest: None
Indochinese Silvered langur status: Occurrence confirmed, last evidence in 1999 (Ngo Van Tri & Day, 1999)

From June to August 1999, FFI conducted a survey to examine the biological diversity of mammals in the area. In July, one troop of at least 9 individuals (estimated at 14) was seen moving quietly in the canopy near the Thac Mai Waterfall (11°06'05"N / 107°25'47"E) (Ngo Van Tri & Day, 1999).

Tay Ninh Province
Special use forest: Partly included in Lo Go Sa Mat Nature Reserve
Indochinese Silvered langur status: Provisional occurrence, last report in 1999 (interview, specimen)(Le Trong Trai, pers. comm., 2000).

One infant was reported to have been collected in Nui Ba Den by A. Morice (1875) (museum unknown). One immature male (museum unknown) collected by J. Delacour and W. P. Lowe between 1926 and

1927 is listed from Tay Ninh (Thomas, 1929). Currently, only 41,000 ha of forest remain in Tay Ninh, of which 24,000 ha are in deplorable condition. The forest in Nui Ba Den Nature Reserve is very small and isolated by human settlements (Ngo Van Tri, pers. comm. 2000).

However, Indochinese Silvered langurs were recently recorded in the province close to the Cambodian border during a BirdLife and FIPI survey. Indeed, one Indochinese Silvered langur tail was seen in Lo Go Sa Mat Nature Reserve in December 1999 (Le Trong Trai, pers. comm. 2000). The species was reported to still be present in the nature reserve by local informants.

Hon Chong Cultural and Historical Site and vicinity (KIEN GIANG)
Special use forest: Partly included in Cultural and Historical Site
Indochinese Silvered langur status: Occurrence confirmed, last evidence in 2001 (Ngo Van Tri, pers. comm. 2001)

In this area probably only two groups occur (one group with 19 individuals inside the protected area and one group with 12 indivuduals outside), which are isolated due to 10 km of agricultural land with rice fields. (Hoang Minh Duc, pers. comm. 2000).

A troop of 18 individuals was seen close to the Hang Pagoda in August 2001 (Ngo Van Tri, pers. comm., 2001). The area is predominantly karst limestone. Photographs of the animals were obtained.

Kien Luong Proposed Nature Reserve (KIEN GIANG)
Special use forest: Proposed nature reserve
Indochinese Silvered langur status: Occurrence confirmed, last evidence in 2001 (Ngo Van Tri, pers. comm., 2001)

A troop of 9 individuals was seen in the limestone karst area of the proposed nature reserve (Ngo Van Tri, pers. comm. 2001).

Locations where Indochinese Silvered langurs are believed to be extinct

Ngoc Hien District (CA MAU)

One infant female skin (IEBR 1530/M143) was collected in April 1977 by Truong Minh Hoat in Tan An locality. No records have been made since and the species is believed to be locally extinct.

Discussed records

Tuyen Hoa District (QUANG BINH) ca. 17°53'N / 106°02'E

Dao Van Tien (undated MS), by identifying *T. cristatus caudalis* from Tuyen Hoa District (Quang Binh Province) at 17°53'N of latitude, would extend the range of the species very far to the North. This report is disputed by Fooden (1996). Such an occurrence would constitute evidence of parapatry of the species with *T. crepusculus*. This issue deserves further investigation.

3.6.5   Status

The Indochinese Silvered langur is widespread but is a very rare species in most of its range.

The apparent extremely low Indochinese Silvered langur density in Vietnam suggests that the population has been seriously reduced due to human pressure. Reports are very scarce, and only few sightings are documented in the last fifty years, though this conclusion may also be the result of the limited number of surveys conducted in the range of the species.

There are only 7 localities where the occurrence of the species was confirmed, and 7 localities are known with provisional occurence.

Information from the further 4 localities is 25 to 90 years old and the current occurrence very questionable.

In Lao PDR, Indochinese Silvered langur is widespread and found in several areas, but no large continuous area is confirmed to support high populations (Duckworth *et al.*, 1999). The species is probably not very numerous and is locally considered **"At Risk"**.

Very little is known about the status of Silvererd langur in Cambodia. The species seems to be found in very localised areas of riverine forest surrounded by evergreen or dry deciduous forest (Long *et al.*, 2000b; Long *et al.*, 2000c).

The Indochinese Silvered langur is listed with the name *T. cristatus margarita*, in the *IUCN Red List Category* as **"Lower Risk: near threatened" LR** 2nt by Eudey (1996/1997), and with the uncommon scientific name *Trachypithecus villosus* in the *IUCN Red List of Threatened Species* (Hilton-Taylor, 2000) and classified as **"Data Deficient"**.

Pham Nhat *et al.* (1998) proposed the status **"Vulnerable"** for Vietnam.

This status does not reflect the current situation and should be upgraded for Vietnam to **"Critically Endangered"** with the criteria **CR** A1cd, A2cd, C2a.

3.6.6   Recommendations for conservation in Vietnam

3.6.6.1   Conduct further field status surveys

Further surveys are urgently needed to improve knowledge of the status and distribution of the Indochinese Silvered langur in Vietnam. Indeed, the estimated range of the species is composed of large areas where very little information is available about the status of the environment. Priority should be given to Binh Phuoc and Tay Ninh Provinces, in areas adjacent to Cambodia and to the Da Lat Plateau area, in northern Lam Dong Province and southern Dak Lak Province.

3.6.6.2   Expansion of protected area network

Most of the area of western Dak Lak Province is dominated by a dry deciduous habitat and, therefore, is not suitable to support a large population of primates. Indochinese Silvered langur groups may still inhabit mixed evergreen forest patches on hills and along rivers. Given that most of the records in Vietnam were made in such areas, the expansion of the protected area network should be highly beneficial for local-level conservation of the species. The expansion of the boundaries of Yok Don National Park to the north, in Ea Sup District, and to the south, in Cu Jut District, as proposed by FFI in 1999, will include two areas where the occurrence of the species was recently confirmed.

4.0. Douc langurs

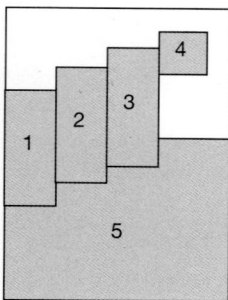

1. Red-shanked douc langur; ad. male. *T. Nadler*

2. Black-shanked douc langur; ad. male. *T. Nadler*

3. Grey-shanked douc langur; ad. male. *T. Nadler*

4. Red-shanked douc langur; infant male, nine month old. *T. Nadler*

5. Red-shanked douc langur; ad. male. *T. Nadler*

4. Douc langurs

4.0 An introduction to the douc langurs (genus: *Pygathrix*)

4.0.1 Taxonomy

Four Asian colobine genera: *Pygathrix, Rhinopithecus, Nasalis* and *Simias* with special nasal pecularities and a high intermembral index are usually combined in the "Odd-Nosed Group".

Taxa currently recognised as belonging to *Pygathrix* were placed under several other genera until Pocock's (1934) revision of genus-level taxonomy. This name was given by Geoffroy St. Hillaire (1812).

The Red-shanked douc langur, *Pygathrix nemaeus*, from northern Vietnam and Lao PDR was described by Linnaeus in 1771. Milne-Edwards (1871) described the southern form, the Black-shanked douc langur, *Pygathrix nigripes*. Kloss (1926), Pocock (1934) and many subsequent reviewers (Napier, 1985; Groves, 1970; Corbet & Hill, 1992; Davies & Oates, 1994; Jablonski, 1995, 1998) considered *P. nigripes* a subspecies of *P. nemaeus*. Brandon-Jones (1984) recognised two species, based mainly on colouration and zoogeographic distribution.

It was not until 1997 that a third taxa was decribed, as a subspecies of the Red-shanked douc langur *P. nemaues cinereus* (Nadler, 1997a). The Latin adjectival subspecific name *cinereus* should be altered to *cinera* so that its suffix agrees in feminine gender with the generic name *Pygathrix* (Timmins & Duckworth 1999).

Recent studies of genetics (Roos & Nadler, 2001) show a divergence on species level, and craniometric studies show that significant differences exist between the three forms (Harding & Groves, 2001). Groves (2001) recognizes the three forms as distinct species.

However, hybrids are known between Red- and Grey-shanked doucs and between Grey- and Black-shanked doucs. Lippold & Vu Ngoc Thanh (1995b) described observations of animals with intermediate colouration and Nadler (1995b) described museum specimens which also most probably are hybrids.

Another form of douc langur, was described as *P. nemaeus moi* by Kloss (1926), on the basis of two males, three females and two juveniles from Nui Ba (=Nui Lang Bian) and of two males collected in Don Duong (=Dran) in Lam Dong Province. Kloss observed that these specimens showed a reduction in the extent of black area compared with the *nigripes* from southern Vietnam. Examinations of the skin described as *moi* by Kloss shows that his original description does not adequately represent the variation observed in pelage colouration of *nigripes* (Thomas, 1928, Jablonsky, 1995).

Most authors regard *moi* as a synonym of *nigripes* (Groves, 1970; Napier, 1985; Weitzel *et al.*, 1988; Jablonski, 1995;), a view which is now supported by DNA investigation (see 2.3).

4.0.2 Morphology, ecology and behaviour

Douc langurs are large colobines and, compared to the langurs, the body appears more heavy and clumsy. The arms are only slightly shorter than the legs. The tail length is approximately equal to the head/body length. The head has no crest on crown, the hairs are smoothly directed backwards and form a short cape over the neck.

The nose is flat, not upturned but little flaps of skin partially cover the nares from the distal-externam margin. The eyes appear almond-shaped and are set at a slightly inclined angle, the lateral aspect of the eye being slightly higher than the medial. The eye shape imparts an Asiatic expression to the douc's face.

The doucs have two distinctive morphological features of the stomach anatomy which are different to the other colobine species (Caton, 1998). The dentition is typically colobine, but the canines are small and the lower third molar sometimes bears six cusps.

The colouration of the sexes is alike except for a white spot exhibited by males at each anterior angle on a small triangular patch above the root of the tail. The ischial callosities are separate in both sexes. The males are larger than females.

The colouration of the three douc species are clearly different. The infants do not show differences but a strong infant-adult dimorphism in the pelage colour is present in each species. Infants of all three species are born with a reddish-blue face mostly with yellow eye-rings slightly similar to adult *nigripes*. In Grey and Red-shanked doucs the face becomes yellow-orange after 2 years.

Douc langurs are entirely arboreal, moving in the trees quadrupedally and brachiating. Observation at the EPRC show that brachiation is a very common locomotion. In the wild, animals are not seen on the ground (Lippold, 1998). It is suggested by Pham Nhat *et al.* (2000) that *P. nemaeus* prefers to sleep in larger trees (30-35m. in height, 1-2m in diameter) with big branches and a thick canopy, although this information is based only on limited evidence of sleeping sites. However, it also seems the genus can adapt to a relatively heavily disturbed forest, as found during FSZ-surveys (Ha Thang Long, pers. comm.).

Doucs are observed from sea level up to an elevation of 1500m (*P. nigripes* Da Lat Plateau, Eames & Robson, 1993).

Douc langurs are mainly folivorous. In field observations, Lippold (1998) estimated that 75% of the diet of her study sample consisted of small tender leaves. Other foods are buds, fruit, seeds and flowers. The analyses of stomach contents by Pham Nhat (1994a) found that doucs use at least 50 plant species and have a strong preference for *Ficus* parts, including fruit, leaves and buds. No animal parts were found in the stomachs in this study.

Douc group size varies depending on locality. Lippold (1995b) reports sightings of groups of 51 individuals in Kon Cha Rang. However, she never found groups of more than 17 individuals in Son Tra. Lippold (1998), comparing the level of disturbance and the size of the groups, found that the average band size is closely related to the habitat quality and human disturbance. This conclusion is supported by Osgood's (1932) observations in areas previous to the spraying of Agent Orange. In Bach Ma National Park this author reported groups of 30 to 50 individuals. In contrast, Lippold (1995b) found group-sizes of no more than 20 at Bach Ma. Pam Nhat *et al.*, (2000) mention groups in Phong Nha-Ke Bang National Park with 15 to 20 up to 30 individuals.

Current reports show a decreasing number of larger groups in some areas, most probably substantiated with habitat destruction, habitat fragmentation and high hunting pressure.

Ratajszczak *et al.* (1992) reported: "After disturbance such as hunting...larger groups disperse into smaller groups from 2 to 20 individuals, apparently with a big male as a leader".

One long term ecological study was conducted on this genus in 1974 on the population of *P. nemaeus* in Son Tra (Lippold, 1977), and was enriched by several surveys carried out between June 1993 and June 1996 in Son Tra, Pu Mat, Bach Ma and Kon Cha Rang (Lippold, 1998). WWF- Indochina Programme also conducted a monitoring and ecological research project focussing on *P. nemaeus* in the Phong Nha-Ke Bang area (Pham Nhat *et al.*, 2000).

Most douc groups are multi-male and multi-female. The male/female ratio seems to correspond loosely with increased group size. A compilation is given by Pham Nhat *et al.* (2000). Smaller groups (3 to 10 individuals) mostly have a male/female ratio of 1:1.5 to 1:2 larger groups (15 to 35 individuals) a ratio of 1:2.5 to 1:3

Groups are characterized by a linear social hierarchy, and males are dominant over females (Nowak, 1991).

Ruempler (1998) reported that females produced their first infant between 5 and 7 years of age and males between 5 and 8 years of age. Napier (1985) gives the age of sexual maturity for females as 48 months and as 60 months by males.

Daily urine analysis at San Diego Zoo indicated a gestation period of 210 days (Lippold, 1981).

Cycling and pregnant females exhibit a reddened inguinal and perineal area, which is otherwise white. Males also show the same feature in response to a cycling female (Lippold, 1998).

The interbirth interval (in captivity) ranges from 16 to 38 months with an average of 24 months (n=24). The oestrus cycle is 28 to 30 days, the litter size one, twins are very rare (Ruempler, 1998).

In terms of behavioural ecology, very little is known about the genus, however, one behavioural feature which may be of particular importance to the species' survival is the animal's reaction to being hunted (shot at). If the group is alarmed the usual response is a deep-throated 'khuk-khuk' sound from the troop leader or at least adult males, whilst the females make a 'kheek-kheek' alarm sound, before the group flees quickly and quietly through the upper storeys of the canopy. However, according to hunters in Phong Nha-Ke Bang when a group is shot at, the animals do not run away, but instead climb higher in the tree and hide behind big branches or will break off branches and bits of foliage to hide their faces (Pham Nhat *et al.*, 2000). It is not known whether this behaviour is species-specific or whether it is found amongst all douc langurs, although the same author reports a similar phenomenon from observation of one group of Black-shanked douc langurs in 1992. Consequently, *P. nemaeus* used to be the easiest species to hunt in Phong Nha-Ke Bang and such behaviour may account for the species' rapid decline.

4.1   Description

4.1.1   Red-shanked douc langur
Pygathrix nemaeus (Linnaeus, 1771)

The Red-shanked douc langur is one of the most colourful primate species. The Vietnamese name "Vooc nu sac" means the langur with five colours. Back, crown and upper arms are dark silver grey, the belly slightly more bright. The hairs in this regions are "agouti". Each single silver grey hair has about 5 to 7 small black rings. The back of the hands and forearms nearly to the elbow are white; fingers and inside of the hand black. Upperlegs are black, lower legs chestnut, and feet black. There is a large white triangle pubic patch in both sexes, the scrotum is white, penis pink, and the tail white with a thin tassel. Above the root of the tail is a white triangular patch with a whorl of white hairs on the anterior corner.

The face around the eyes and nose is yellow-orange to light brown, around the mouth white, and the side of the head is white to the ears. The axis of the eyes is high (20.9°) which imparts the Asiatic expression. A wide forehead band is black, the face is surrounded by white whiskers as long as 12cm in adult males; a chestnut collar separates the white throat from a black brustband which runs above the grey belly from shoulder to shoulder.

External measurements (adult)

Pygathrix nemaeus		n	mm	Average	Source
Head/Body length					
	male	?	550-630	586	Napier 1985
		?	550-820		Walker 1983
		2	620	620	EPRC
	female	?	597		Napier 1985
		?	597-630		Walker 1983
Tail length					
	male	?	600-735	680	Napier 1985
		2	560-670	615	EPRC
	female	?	597		Napier 1985
		6	440- 575	525	San Diego Zoo
Weight			kg		
	male	?	10.9		Fleagle 1988
		?	9-15		Walker 1983
		3	8.6-11	9.8	EPRC
	female	?	8.2		Fleagle 1988
		?	7-10		Walker 1983
		7	6.84-8.3	7.3	San Diego Zoo
		4	(5.2) 6.6-8.4	(6.5)7.5	EPRC

Pygathrix nigripes		n	mm	Average	Source
Head/Body length					
	male	1	500	500	EPRC
	female	1	560	560	EPRC
Tail length					
	male	1	690	690	EPRC
Weight			kg		
	male	1	5.3	5.3	EPRC
	female	3	5.0- 5.3	5. 4	EPRC

Pygathrix cinerea		n	mm	Average	Source
Head/Body length					
	male	3	620-640	630	EPRC
	male	2	560-580	570	EPRC
Tail length					
	male	2	640-650	645	EPRC
	female	2	590-640	615	EPRC
Weight			kg		
	male	3	10.7-12.4	11.5	EPRC
	female	2	7.5-9.4	8.45	EPRC

4.1.2   Black-shanked douc langur
Pygathrix nigripes (Milne-Edwards, 1871)

The grey tone on the back is darker than in Red-shanked doucs but the belly is more bright. The hairs are also "agouti". Upper and forearms are very dark grey, nearly black, and the hands are black. The upper and lower legs are black, and the feet black. The white pubic patch is smaller than in Red- and Grey-shanked doucs, the scrotum is blue and the penis red. The tail is white (longer than in the other species) and the tassel is very thin or absent. The white whorls on the patch above the tail in males are not so clear as in Red- and Grey-shanked doucs.

The face is bluish with large yellow eye-rings. The axis of the eyes is lower than in both other species (11.2°) which yields a different facial expression. The black forehead band runs along the side of the head down to the shoulders. In front of the ears there is a black triangle, and the ears are inside the black area. The white throat is surrounded by a chestnut collar which comes close to the black band running from shoulder to shoulder. The whiskers are short and thin.

4.1.3   Grey-shanked douc langur
Pygathrix cinerea Nadler, 1997

The crown, back, arms and legs are nearly the same light grey colour with "agouti" hairs. The fingers and feet are black; the belly very light grey, sometimes nearly white. The tail is like that of the Red-shanked douc, with a thin tassel.

The pubic patch, colour of scrotum and penis are similar to those of Red-shanked doucs.

The colouration of the face, throat and collar is like the Red-shanked douc, although the whiskers are shorter and not so dense. The axis of the eyes (14.6°) is between the two other species. The black forehead band is commonly a bit smaller than in Red-shanked doucs.

4.2   Distribution

The douc langurs are restricted to Vietnam, Lao PDR and Cambodia, east of the Mekong River (Corbet & Hill, 1992; Fooden, 1996). *P. nemaeus*, the northern form, extends from about 18°40'N latitude in Lao PDR and 19°30'N in Vietnam southwards and is replaced by *P. cinerea* in the central provinces of Vietnam, and *P. nigripes* in south Vietnam and east Cambodia. Our current knowledge indicates that each taxon is at least parapatric with the others, and areas with sympatric distribution, where interbreeding may occur, are limited (Vu Ngoc Thanh, pers. comm., 2000; Ha Thang Long, pers. comm., 2000).

Distribution in Lao PDR

Douc langurs are widely distributed in central and southern Lao PDR, south of 18°40'N (Nam Kading NBCA and Nam Theun Extension PNBCA), from the Mekong River to at least 1.600m a.s.l. (Timmins & Duckworth, 1999). The species is usually absent from degraded, fragmented and largely deciduous forests. The species remains common in some remote areas, but in fragmented areas populations have declined rapidly (Duckworth *et al.*, 1999)

All douc langurs in Lao PDR belong to *P. nemaeus*. The occurrence of *P. cinerea* and *P. nigripes* has never been confirmed. However, in Dong Ampham NBCA and Phou Kathong PNBCA (extreme south) some animals show reduced red suggesting some intergradation with another form (Davidson *et al.*, 1997; Timmins & Duckworth, 1999).

Distribution in Cambodia

Doucs may occur only east of the Mekong River (Corbet & Hill, 1992). Recent observations report the occurrence of *P. nigripes*. Sightings of the species were obtained in Snuol Wildlife Sanctuary (J. Walston, pers. comm., 2000), in southern Ratanakiri province (Timmins & Mey Soriyun, 1999) and in northeastern and southern Mondulkiri Province close to Dak Lak Province border in Vietnam (Long *et al.*, 2000c.; Walston *et al.*, 2001). The identification of *P. nigripes* by Long *et al.*, 2000c was supported by DNA analysis (C. Roos, pers. comm., 2000) conducted on two skin samples taken from dead animals found in traps set for large carnivores.

One photo taken by Redford 1999 (unpubl.) in north-east Ratanakiri Province, close to the border of Lao PDR and Vietnam (14°19'N/107°17'E) appears to be that of a Grey-shanked douc langur.

Distribution in Vietnam

The ranges of the three species are aligned from north to south in the following order: Red-, Grey- and Black-shanked doucs. However, the data currently available about their distribution does not allow an exact delineation of the boundaries among the species. There is also confusion in former observations - the Grey-shanked form was first described only in 1997 - and hybrids are also known at some locations. For this reason, the species inhabiting each locality is noted, as well as the diagnostic evidence. Douc langurs are widely distributed from Nghe An Province to the Mekong Delta. They are very scarce or absent in the largely deciduous forests of western Dak Lak and southern Gia Lai Provinces and restricted to pockets of mixed evergreen forest. *P. nemaeus* is confirmed from 19°02'N (Pu Mat National Park) to 14°33'N (Kon Ka Kinh Nature Reserve). *P. nigripes* is confirmed from 14°33'N (Kon Ka Kinh Nature Reserve) to 11°26'N (Cat Tien National Park). *P. cinerea* occurs between the other species; there are provisional reports of its occurrence from about 16°N (Hien District, Quang Nam Province) to 14°25'N (K' Bang and Mang Yang Districts, Gia Lai Province), and confirmed reports from 15°25'N (Tien Phuoc District, Quang Nam Province) to 14°25'N (K' Bang and Mang Yang Districts, Gia Lai Province).

Knowledge about the range of the three species still requires clarification. Currently, no zoogeographic boundaries can be determined with any precision. Although *P. nemaeus* is found as far south as the northern Gia Lai Province, its occurrence between there and Da Nang Province has not been recorded. Between the two areas, only *P. cinerea* is recorded. In the Mom Ray area (Kon Tum Province) *nigripes* and *nemaeus* are provisionally reported. Hybrids may also inhabit the region. In the northern Gia Lai Province (Kon Ka Kinh and Kong Cha Rang Nature Reserves), the three taxa live probably in close proximity. In Kong Cha Rang Nature Reserve, *P. nigripes* has been provisionally reported and specimens belonging to *nemaeus* have also been found. In Kon Ka Kinh Nature Reserve, even further south, only *P. nemaeus* is reported. Furthermore, *P. cinerea* was reported to have been caught between Kong Cha Rang and in Kon Ka Kinh Nature Reserves. Lippold (1995a,b) claimed that *nemaeus* and *nigripes* were sympatric in the northern area of Gia Lai Province but this has not been confirmed.

DISTRIBUTION OF RED-SHANKED DOUC LANGUR *(Pygathrix nemaeus)* IN VIETNAM
BẢN ĐỒ PHÂN BỐ LOÀI CHÀ VÁ CHÂN NÂU Ở VIỆT NAM

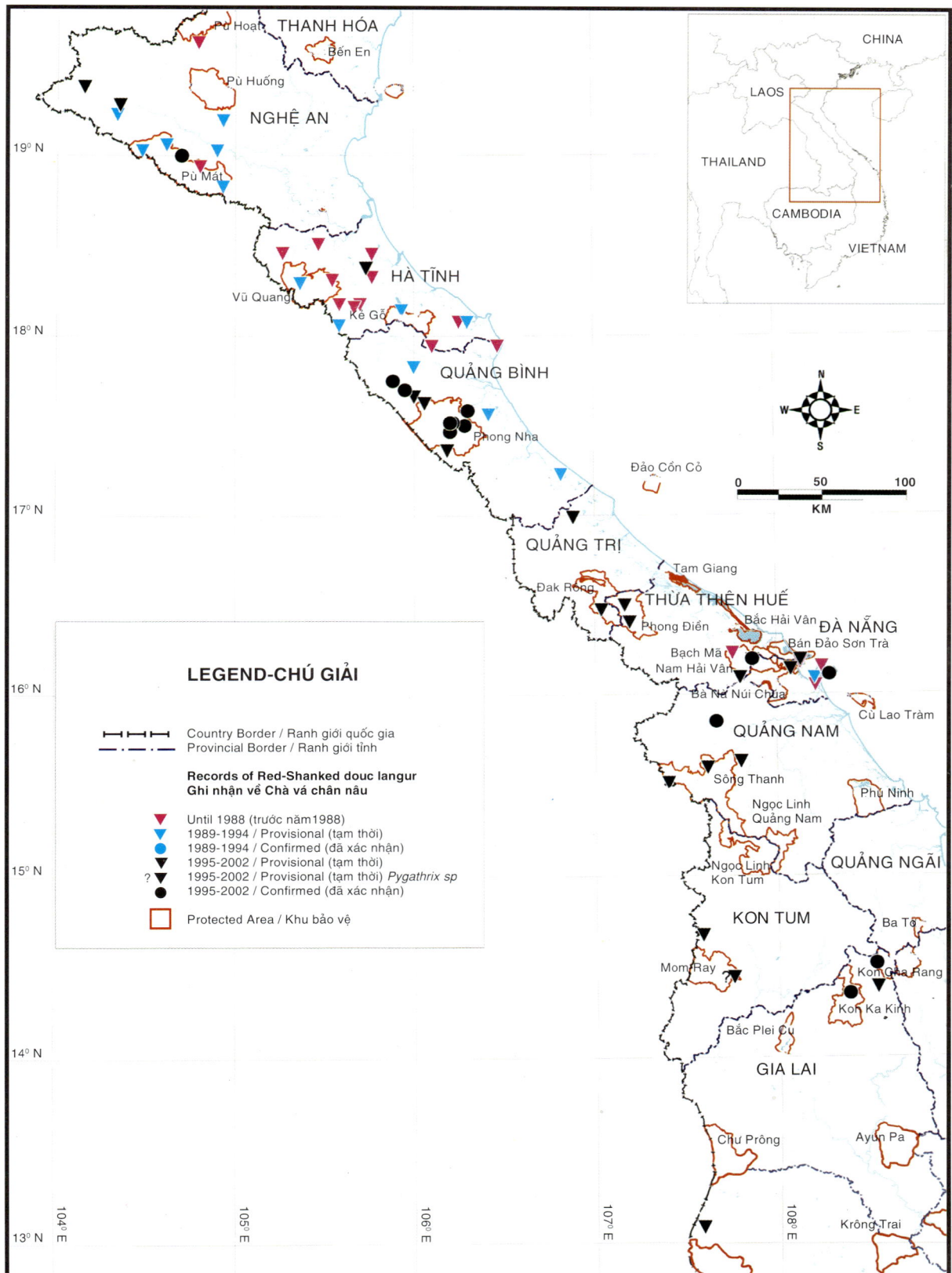

LEGEND-CHÚ GIẢI

⊢⊢⊢	Country Border / Ranh giới quốc gia
—·—·—	Provincial Border / Ranh giới tỉnh

Records of Red-Shanked douc langur
Ghi nhận về Chà vá chân nâu

▼ Until 1988 (trước năm 1988)
▼ 1989-1994 / Provisional (tạm thời)
● 1989-1994 / Confirmed (đã xác nhận)
▼ 1995-2002 / Provisional (tạm thời)
? ▼ 1995-2002 / Provisional (tạm thời) *Pygathrix sp*
● 1995-2002 / Confirmed (đã xác nhận)

☐ Protected Area / Khu bảo vệ

DISTRIBUTION OF GREY-SHANKED DOUC LANGUR *(Pygathrix cinerea)* IN VIETNAM
BẢN ĐỒ PHÂN BỐ LOÀI CHÀ VÁ CHÂN XÁM Ở VIỆT NAM

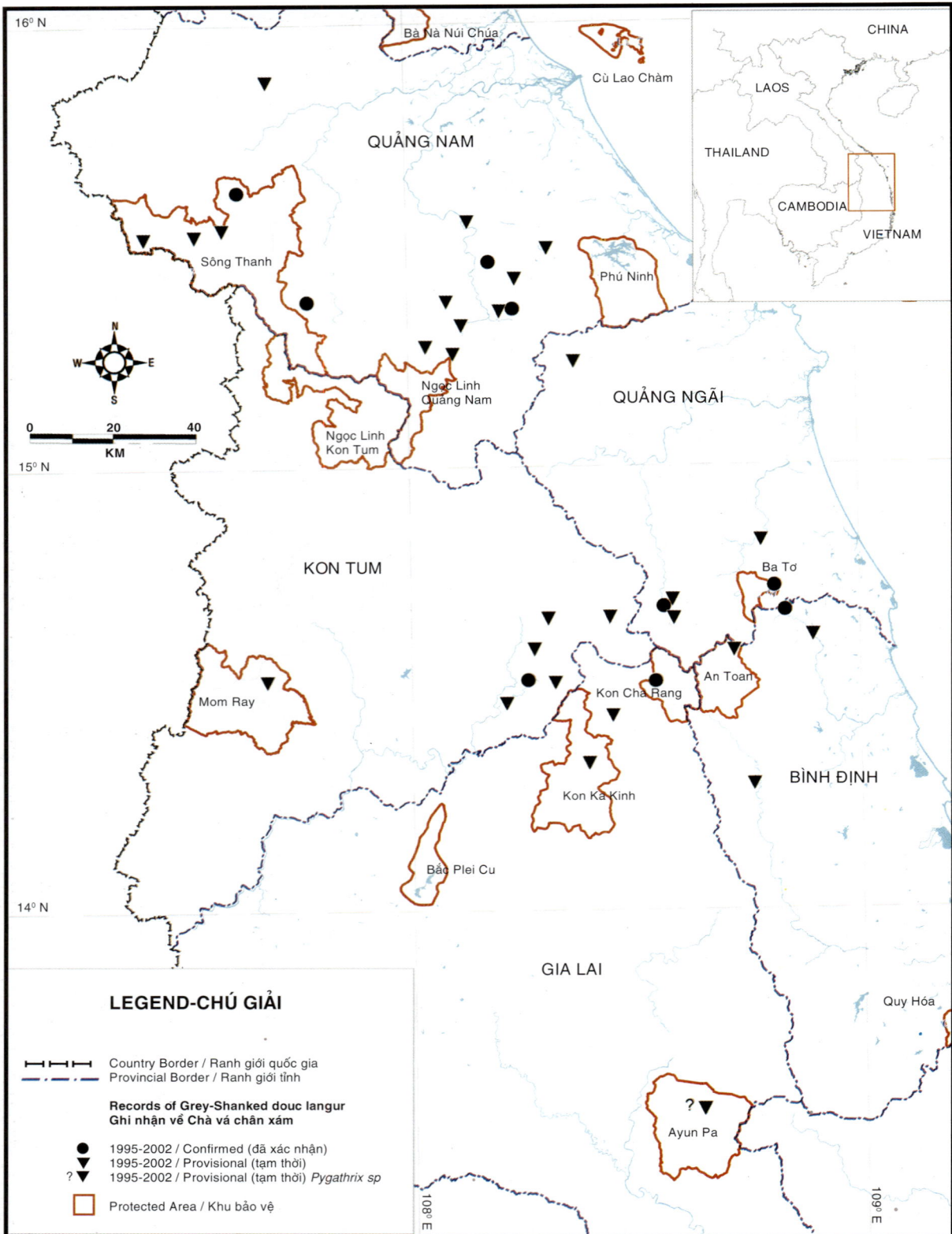

LEGEND-CHÚ GIẢI

Country Border / Ranh giới quốc gia
Provincial Border / Ranh giới tỉnh

Records of Grey-Shanked douc langur
Ghi nhận về Chà vá chân xám

● 1995-2002 / Confirmed (đã xác nhận)
▼ 1995-2002 / Provisional (tạm thời)
? ▼ 1995-2002 / Provisional (tạm thời) *Pygathrix sp*

☐ Protected Area / Khu bảo vệ

DISTRIBUTION OF BLACK-SHANKED DOUC LANGUR *(Pygathrix nigripes)* IN VIETNAM
BẢN ĐỒ PHÂN BỐ LOÀI CHÀ VÁ CHÂN ĐEN Ở VIỆT NAM

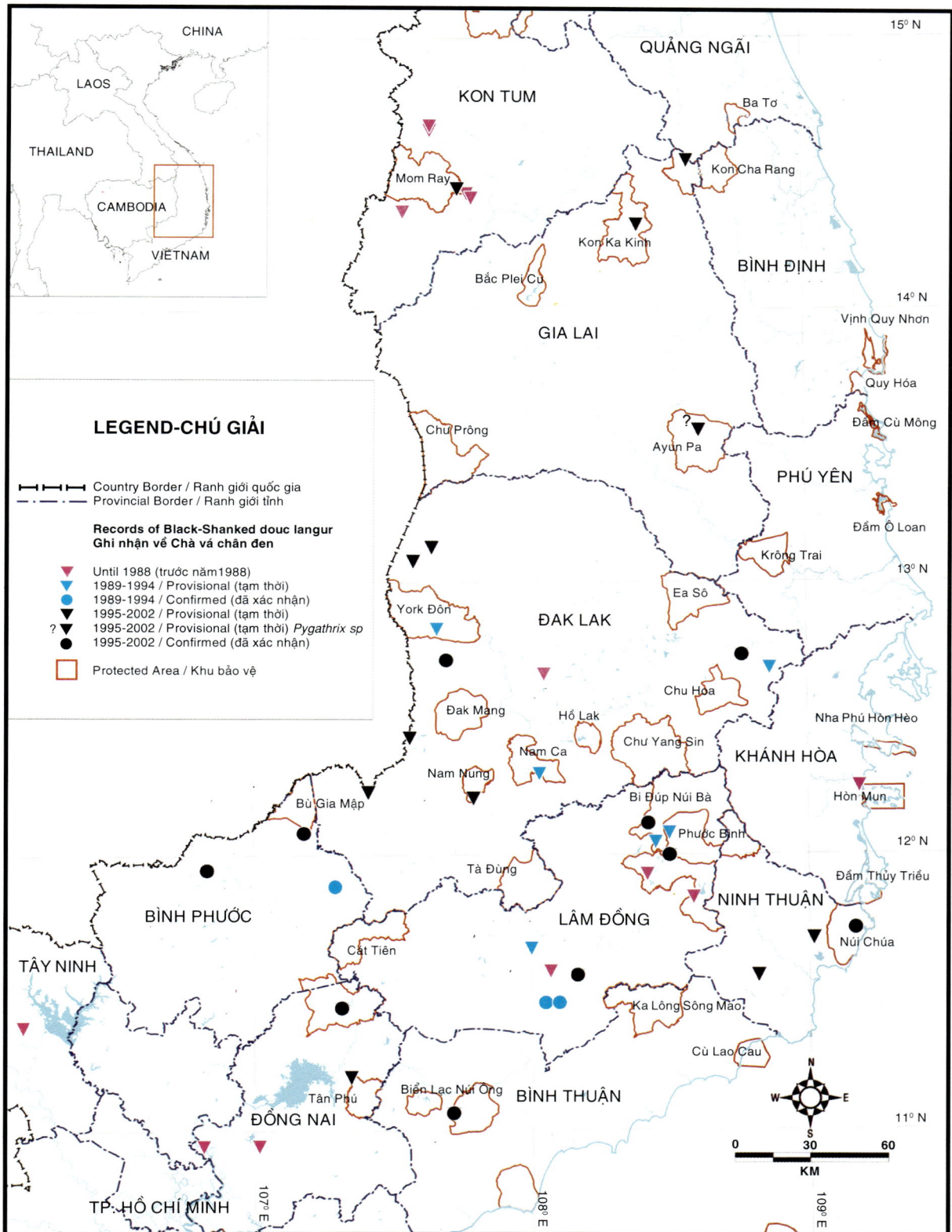

CHINA

LAOS

THAILAND

CAMBODIA

VIETNAM

15° N

QUẢNG NGÃI

KON TUM

Ba Tơ

Kon Cha Rang

Mom Ray

Kon Ka Kinh

BÌNH ĐỊNH

Bắc Plei Củ

14° N

Vịnh Quy Nhơn

GIA LAI

Quy Hóa

Chư Prông

Đầm Cù Mông

?

Ayun Pa

PHÚ YÊN

LEGEND-CHÚ GIẢI

Đầm Ô Loan

Country Border / Ranh giới quốc gia
Provincial Border / Ranh giới tỉnh

Krông Trai

13° N

**Records of Black-Shanked douc langur
Ghi nhận về Chà vá chân đen**

Ea Sô

▼ Until 1988 (trước năm1988)
▼ 1989-1994 / Provisional (tạm thời)
● 1989-1994 / Confirmed (đã xác nhận)
▼ 1995-2002 / Provisional (tạm thời)
? ▼ 1995-2002 / Provisional (tạm thời) *Pygathrix sp*
● 1995-2002 / Confirmed (đã xác nhận)

York Đôn

ĐAK LAK

Chu Hoa

Nha Phú Hòn Hèo

☐ Protected Area / Khu bảo vệ

Đak Mang

Hồ Lak

Nam Ca

Chư Yang Sin

KHÁNH HÒA

Nam Nung

Hòn Mun

Bù Gia Mập

Bi Đúp Núi Bà

Phước Bình

12° N

Tà Đùng

Đầm Thủy Triều

BÌNH PHƯỚC

NINH THUẬN

Cát Tiên

TÂY NINH

LÂM ĐỒNG

Núi Chúa

Ka Lông Sông Mao

Cù Lao Cau

Tân Phú

Biển Lạc Núi Ông

BÌNH THUẬN

ĐỒNG NAI

11° N

TP. HỒ CHÍ MINH

0   30   60
KM

107° E   108° E   109° E

4.3   Douc langur records in Vietnam

Pu Mat National Park and vicinity (NGHE AN)
Special use forest: partly included in Pu Mat National Park
Douc status: Occurrence confirmed, last evidence in 1995 (Vu Ngoc Thanh, pers. comm., 2000)
Species recorded: *P. nemaeus* (sighting and specimen).

Several surveys report the occurrence of douc langurs in Pu Mat National Park with the initial report being Ratajszczak *et al.* (1990). Interviews conducted by this survey team in Cao Veu village reported that the douc was locally common in Anh Son District. In addition, one stuffed animal was seen. This specimen was reported to have been shot from a group of at least 30 individuals in June 1989 near the Lao border of Tuong Duong District. In December 1992, two groups of 9 and 12 individuals were observed on two occasions in Con Cuong District by Pham Nhat (pers. comm., 2000). In May 1995, the same author observed only one group of 4 individuals in the same location. One skull of an adult female (FCXM 006) was collected in October 1994 in Cao Veu village (Fooden, 1996). Lippold (1995a) reports sightings of three groups of 25, 31 and 35 individuals "in several areas of the reserve". In April 1999, one caged immature Red-shanked douc was seen in the upper Khe Thoi River by an SFNC team, although the capture location was not identified and could have been in adjacent Lao PDR as the Khe Thoi River is a major trading route into Lao PDR (Round, 1999).

The surveys by SFNC in both 1998 and 1999 failed to encounter douc langurs. Hunters reported that the species has not been seen recently. A survey focusing on *P. nemaeus* was conducted in the most likely area still to contain the species, but no sightings occurred and local people reported a complete lack of sightings in the previous two years (John, 1999; B. Long, pers. comm.). This suggests that the species may have been largely, if not completely, extirpated from Pu Mat in the last five years.

Huong Son District (HA TINH)
Special use forest: None
Douc status: Occurrence confirmed, last evidence in 2000 (Osborn & Furey, pers. comm.)
Species recorded: *P. nemaeus* (sighting and specimen)

One female skull (FCXM 004) was collected in June 1985 by Nguyen Van Dung at the Son Kim locality. In this district from May to August 1985, Pham Nhat (pers. comm., 2000) observed groups of 4 to 7 Red-shanked douc langurs on five occasions. Ratajsczcak *et al.* (1990) reported that an animal was shot in the district in 1988 on the basis of interviews with local hunters. In June 1990, Rozenddal (1990) reported the occurrence of the species based on interviews and a specimen seen in the headquarters of Forest Enterprise. Individuals were seen in February 2000 at 18°22'34.3"N / 105°12'14.6"E during a survey conducted by a SEE team (Osborn & Furey, pers. comm.).

Vu Quang National Park (HA TINH)
Special use forest: National park
Douc status: Occurrence confirmed, last evidence in 2000 (Pham Nhat *et al.*, 2000)
Species recorded: *P. nemaeus* (sighting and specimen)

Several museum specimens of Red-shanked douc langur come from Huong Khe District. One adult male skin (ZMVNU 104) was collected in April 1961 by Nong Ke Nhau in the Trai Tru locality. Dao (1985) reports that a female (museum unknown) was collected in the same place in November 1964.

One infant skull (IEBR 43), one adult male (IEBR 432), one juvenile skull (IEBR 437) and one female mandible (IEBR 497) were collected by Thai Khac Zieu in January 1964 in the Truong Bat locality (=Truong Hat). One female skin (IEBR 433/M21) was collected in January 1964 (unknown collector) and reported from Huong Khe District. One juvenile skull (IEBR 43/106) and one adult female skull (IEBR 43/107, date and collector unknown) are also reported from the district.

A survey by a Hanoi University team observed one group of about 5 individuals in August 1974 in Huong Dien Commune (Vu Ngoc Thanh, pers. comm.). Pham Nhat observed groups of 7 and 8 individuals respectively in July and August 1992 in the forests of Huong Khe District (Pham Nhat, pers. comm. 2000).

Sightings were reported in the nature reserve during a survey for endemic pheasants, conducted in June and July 1994 by NWF, IUCN, WWF and BirdLife (Lambert *et al.,* 1994). In July and August 1997, doucs were recorded, based on interviews by scientists from VRTC (VRTC 1997). In December 2000, one sighting was recorded in the Hung Lau area during a monitoring programme (Pham Nhat *et al.,* 2000)

According to local hunters, the population has rapidly decreased in the last five years. For example, over a 6-day survey in August 1995, Pham Nhat found no sign of this animal (Pham Nhat, pers. comm., 2000).

Ke Go Nature Reserve (HA TINH)
Special use forest: Nature reserve
Douc status: Provisional occurrence, last report in 1995 (Le Trong Trai *et al.,* 1996)
Species recorded: *P. nemaeus* (interview and specimen)

Two specimens of Red-shanked douc langur, one female skin (IEBR 42/440) and one adult female skull (IEBR 441/49) were collected in January 1964 by Nguyen Lien, reportedly from the Ky Son locality (Ky Anh District).

Ratajsczcak *et al.* (1990), reported from interviews with local hunters that douc langurs can be found in primary and secondary forests in the vicinity of the southern and western shores of Ke Go lake. This occurrence is also recorded by BirdLife and FIPI, on the basis of interviews conducted during a field survey, carried out between April and August 1995 for the investment plan of Ho Ke Go Nature Reserve (Le Trong Trai *et al.,*1996).

Human pressure on wildlife and its natural habitat is very high in Ke Go and it is unlikely that the nature reserve contains a significant douc langur population.

Phong Nha-Ke Bang National Park and vicinity (QUANG BINH)
Special use forest: Partly included in Phong Nha National Park
Douc status: Occurrence confirmed, last evidence in 2002 (Le Khac Quyet, 2002)
Species recorded: *P. nemaeus* (sighting)

In 1995, the area was surveyed with a specific focus on the Hatinh Langur (Pham Nhat *et al.,* 1996a). Three groups of Red-shanked douc langur were observed. One group (number of individuals unknown) was seen in Truong Hoa Commune (northwest of the proposed park) and two other groups were seen in Thung Nui Treo and Phong Nha villages, Son Trach Commune (east of the park) with respectively 10 and 12-15 individuals. In addition, one captive animal was observed in Ban Ruc (village), Truong Hoa Commune (northwest of the park).

Only one group of 8 Red-shanked doucs, without young animals, was observed on the route from Eo Cap to Vooc (17°32'N / 106°12'E) in July 1998 (Nguyen Xuan Dang *et al.*, 1998). Other reports are based on interviews with local hunters of Ban Ruc (Thuong Hoa Commune) and Tang Hoa (Hoa Son Commune).

According to Pham Nhat & Nguyen Xuan Dang (1999), it is likely that Phong Nha possesses one of the largest populations of Red-shanked doucs in Vietnam. However, from three surveys in this area, totalling 30 days in the field, Vu Ngoc Thanh (pers. comm., 2000) found no signs of this species. In addition, only one sighting of a small group without juveniles was reported in over four months of intensive field survey by FFI teams from July to October 1998 (Timmins *et al.*, 1999). The apparent frequency of sightings of the species by previous researchers suggests a major population decline in the Phong Nha-Ke Bang area (Timmins *et al.*, 1999).

Pham Nhat *et al.* (2000) observed two groups of Red-shanked douc langurs on 22 and 23 April, 2000 in the Khe Rong area; groups of Douc langurs on four separate occasions in the Hung Lau-Da Ban area, although it is probable that these were not all separate groups; a group of 22 animals was observed twice on 22 and 26 July, 2000 and a group of 7-8 individuals was observed on 22 July. Furthermore, a group of more than 10 individuals was encountered on 12 December, 2000. During another observation the number of individuals could not be identified due to heavy rain. Also one group of 10-15 Red-shanked douc langurs was observed on 23 July, 2000 (Pham Nhat *et al.*, 2000). The reports concurs with Timmins *et al.* (1998) that the population appears to be in decline due to human impacts on their habitat such as snare setting and tree-oil extraction.

Records cited in Le Xuan Canh *et al.* (1997b) are not considered as admissible here. From 38 days in the field, the authors reports more sightings of primates than from any other survey in Phong Nha, and they do not provide justifications for these findings, nor explain how species and numbers were established.

Quang Trach District (QUANG BINH)
Special use forest: None
Douc status: Unknown, last report in 1960 (museum specimen)
Species recorded: *P. nemaeus* (specimen)

One juvenile male Red-shanked douc langur (ZMVNU 101) was collected in April 1960 by Bach Thu Chon in Deo Ngang mountains.

Vinh Linh District (QUANG TRI)
Special use forest: None
Douc status: Provisional occurrence, last report in 1997 (Nadler, 1997c)
Species recorded: *P. nemaeus* (living animal)

In May 1997, the EPRC received one adult male which had been hunted and confiscated in this area.

Dakrong Nature Reserve (QUANG TRI) and Phong Dien Nature Reserve (THUA THIEN-HUE)
Special use forest: Nature reserve
Douc status: Provisional occurrence, last report in 1998 (Le Trong Trai & Richardson, 1999a)
Species recorded: *P. nemaeus* (interview and specimen)

This area possesses the best extent of lowland forest in Central Vietnam. During the feasibility study to establish two nature reserves there, carried out by BirdLife and FIPI in June and July 1998, Red-shanked douc langurs were reported in interviews and one specimen identified (Le Trong Trai & Richardson, 1999a).

Bach Ma National Park (THUA THIEN-HUE)
Special use forest: National park
Douc status: Occurrence confirmed, last evidence in 1997 (Huynh Van Keo & Van Ngoc Thinh, 1998)
Species recorded: *P. nemaeus* (sightings)

One group of 3 to 4 Red-shanked douc langurs was seen by Eames & Robson (1993) in January 1990 in secondary forest during a primate survey in southern Vietnam. The same team observed one single male in February 1990 in primary forest at an altitude of 700 m. Lippold (1995a) reported sightings of three groups of 5, 15 and 20 individuals in primary and secondary forest during surveys conducted in 1994 and 1995. Huynh Van Keo (1998) reported two sightings by park forest guards before September 1994 and in August 1996 of groups with about 5 and 10-15 animals. In May 1997, Matsumura discovered a group of about 5 to 7 individuals (Huynh Van Keo & Van Ngo Thinh, 1998).

Due to the high hunting pressure in Bach Ma National Park, douc langur populations have been considerably reduced (Vu Ngoc Thanh, pers. comm.).

Nam Dong District (THUA THIEN-HUE)
Special use forest: None
Douc status: Provisional occurrence, last report in 2000 (Ha Thang Long, 2000)
Species recorded: *P. nemaeus* (interview)

During a FZS-survey conducted in October 2000 in the district south-west of Bach Ma National Park locals reported the occurrence of Red-shanked douc langurs. This species was known and reported as common until about 1998. The most recent sightings were reported from December 1999. The largest group was seen in 1990 with about 50 to 60 individulas around Khe Cha Mon (Thuong Lo village). The group size is now quite reduced and includes no more than 10 individuals.

Ba Na-Nui Chua Nature Reserve (DA NANG)
Special use forest: Nature reserve
Douc status: Provisional occurrence, last report in 1999 (Vu Ngoc Thanh, pers. comm.)
Species recorded: *P. nemaeus* (interview and specimen)

Doucs were reported in 1994 by local hunters to be common in the western and eastern parts of Ba Na during the feasibility study to establish Ba Na-Nui Chua Nature Reserve (Anon., 1994f).

One living Red-shanked douc langur, confiscated from the FPD station of the nature reserve, was seen in April 1999 by Vu Ngoc Thanh (pers. comm., 2000)

Phu Loc District (THUA THIEN-HUE) and Hoa Vang District (DA NANG)
Special use forest: None
Douc status: Unknown, last report before 1932 (Osgood, 1932)
Species recorded: *P. nemaeus* (sighting and specimen)

One adult male (BMNH 1926.10.4.4) and one adult female (BMNH 1926.10.4.5) were collected by H. Stevens, J. Delacour and W. P. Lowe in February 1926 from Deo Hai Van (=Col des Nuages). Van Peenen *et al.* (1969) cited a specimen also reported from Deo Hai Van (=Col des nuages) and kept in the USNM. These specimens belong to *P. nemaeus*. Osgood (1932) claimed to have seen groups of 30 to 50 individuals in this forest.

Ban Dao Son Tra Nature Reserve (DA NANG)

Special use forest: Nature reserve
Douc status: Occurrence confirmed, last evidence in 1995 (Lippold, 1995b)
Species recorded: *P. nemaeus* (sighting and specimen)

Several museum specimens of Red-shanked douc langurs were collected in this area. One male foetus kept at the MNHN was collected in 1837 by J. F.T. Eydoux (reported from Da Nang vicinity). A specimen, kept in the USNM is listed by Lippold (1977). One adult female skull (FCXM 002) was collected in May 1984 (collector unknown). One female skin (FCXM unnumbered) was collected in November 1989 (collector unknown).

Van Peenen *et al.* (1969) observed several groups from 200m a.s.l. to the top of Son Tra in September 1967. Between May 1967 and May 1968, Gochfeld (1974) reported two groups of 5 and 6 individuals at an altitude of 696m a.s.l. (Lippold, 1977). From June to August 1974, Lippold (1977) observed three groups of 8, 9 and 11 individuals. Pham Nhat (1994) saw one group of 10 individuals in August 1988 and one of 7 in December 1988. Lippold (1995b) reports two groups of 6 and 17 individuals between 1994 and 1995.

Son Tra has been visited many times by survey teams. It is the only place where a long-term douc field survey has been conducted (Lippold, 1977, 1995). Although the forest still has scars left from the defoliant sprayed during the American war and none of the forest remains undisturbed, douc langurs continue to inhabit the area. Vu Ngoc Thanh (pers. comm.) estimates that only 3 groups of about 15 individuals each remain on the peninsula. It seems that the population is now very small or extirpated. Pham Nhat did not observe any doucs during a visit in October 1998 (Pham Nhat, pers. comm.).

Hien District (QUANG NAM)

Special use forest: None
Douc status: Provisional occurrence, last report in 2000 (Ha Thang Long, 2000)
Species recorded: *P. nemaeus* (interview, specimen) *P. cinerea* (interview)

Grey-shanked douc langurs were recorded, based on interviews in May 1997 by Vu Ngoc Thanh and Lippold (Vu Ngoc Thanh, pers. comm.). During a FZS survey conducted in 2000 it was found that most of the locals and hunters know the two forms of doucs (Ha Thang Long, 2000). Grey-shanked doucs had been frequently observed in the southern part, close to Nam Giang district while Red-shanked doucs obviously occurred frequently in the northern part, close to Phu Loc District, Thua Thien-Hue Province.

It is quite likely that the borderline between the distribution of the Red-shanked and Grey-shanked doucs runs roughly along the border of Quang Nam and Thua Thien-Hue Provinces.

There are also reports of the Grey-shanked douc langur close to the border with Lao PDR so, it is quite likely that this species ranges into Lao PDR.

Song Thanh Nature Reserve and Nam Giang District (QUANG NAM)

Special use forest: Nature reserve
Douc Status: Occurrence confirmed, last evidence in 2000 (Ha Thang Long, 2000; Ngo Van Tri, in prep.)
Species recorded: *P. nemaeus* (interview), *P. cinerea* (interview, specimen and living animal)

Giao (1997) first listed *P. nemaeus* in the Song Thanh-Dak Pring area, but its occurrence remained provisional until 2000. Furthermore, one local guide reported the occurrence of one group of *P. nemaeus* in a forested area between Song Thanh NR and Ca Dy SFE in 1999. The presence of doucs were also confirmed by the observation of 15 skulls found (without species identification) in Thanh

My Town and Ro Village, Ca Dy Commune, although hunters could not distinguish between Red and Grey-shanked douc langurs (Ngo Van Tri, in prep.). In contrast, Ha Thang Long (2000) reported that the two douc forms are well known by hunters (Ca Tu minority). According to hunters, the Red-shanked douc is rather rare in the area. But locals from Tabhing, Chaval and Ladee village close to the border with Lao PDR, frequently observe Red-shanked doucs.

Grey-shanked observations are higher around Ca Dy village. Grey-shanked douc langurs were reported from interviews conducted in the area in May 1997 by Vu Ngoc Thanh and Lippold (Vu Ngoc Thanh, pers. comm.).

One ranger report from Nam Giang FPD noted that four Grey-shanked douc langurs had been hunted for research in 1998 around Ca Dy village.

DNA investigation of a skull and a tail, collected during the FZS survey in 2000 at Tabhing village, identified them as belonging to *P. cinerea*.

One Grey-shanked douc langur female was captured by a villager and sold to a policeman who kept it as a pet until it was confiscated by Quang Nam provincial FPD in September 2000 and is now under the care of the EPRC.

Locals confirmed the occurence of *P. cinerea* in Song Thanh Nature Reserve (Ha Thang Long, 2000; Ngo Van Tri, in prep.). This Nature Reserve represents an important conservation status for the Grey-shanked douc langur.

Tien Phuoc District (QUANG NAM)
Special use forest: None
Douc status: Occurrence confirmed, last evidence in 1999 (Vu Ngoc Thanh, pers. comm.)
Species recorded: *P. cinerea* (sighting and living animal)

In May 1999, this area was surveyed with a special focus on the Grey-shanked douc langur. Two groups of *P. cinerea* were seen in the forest and one photo was taken by Vu Ngoc Thanh. In addition, living and dead specimens were seen in Tien Phuoc City market (Vu Ngoc Thanh, pers. comm.).

During a FZS survey in July 2000, based on interviews it was estimated that a population of about 70 to 80 Grey-shanked douc langurs live in the district. The habitat is mostly poor secondary forest.

In January 1998 the EPRC received two confiscated *P. cinerea* from this district.

Phuoc Son District (QUANG NAM)
Special use forest: None
Douc status: Occurrence confirmed, last evidence in 1998 (Vu Ngoc Thanh, pers. comm.)
Species recorded: *P. cinerea* (sighting and specimen)

One group of doucs was observed in May 1998 during a FPD and IEBR survey for tigers. Dead specimens were seen in Phuoc Son City market. (Vu Ngoc Thanh, pers. comm.)

Tra My District (QUANG NAM)
Special use forest: None
Douc status: Occurrence confirmed, last evidence in 2000 (Tordoff *et al.*, 2000)
Species recorded: *P. cinerea* (interview and specimen)

Grey-shanked douc langurs were recorded in Tra My District from interviews by Vu Ngoc Thanh (pers. comm.) during two surveys by the FPD and IEBR for tigers in May 1998 and March 1999. Several skulls were observed in hunter's houses.

From March to June 1999, BirdLife and FIPI conducted a feasibility study to establish Ngoc Linh Nature Reserve (Tordoff *et al.*, 2000a). Doucs were observed on a single occasion, however, the taxonomy of these animals could not be assigned with any certainty to either *cinerea* or *nigripes* (A. Tordoff, pers. comm., 2000). There are no records from the surveys conducted in 1996 and 1998 in Ngoc Linh Nature Reserve (Le Trong Trai & Richardson, 1999b).

From March 1999 to September 1999, J. Harding (Australian National University) conducted a survey of the communes Tra Doc, Tra Bui, Tra Don and Tra Mai and recorded Grey-shanked douc observations from locals. Skulls of Grey-shanked doucs were collected from Tra Bui, Tra Doc and Tra Don (Harding & Groves, 2001).

During a FZS survey in June 2000 three skulls were collected and indentifyied as *cinerea*. Sightings by villagers confirmed that the population is decreasing.

Tra Bong District (QUANG NGAI)
Special use forest: None
Douc status: Provisional occurence, last report in 2000 (Ha Thang Long, 2000)
Species recorded: *P. cinerea* (interview)

During a FZS survey in June 2000, observations of Grey-shanked douc langurs were reported from locals and hunters in Tra Bong Disrict (Ha Thang Long, 2000).

Ba To District (QUANG NGAI)
Special use forest: Partly included in Ba To Cultural and Historical Site
Douc status: Occurrence confirmed, last evidence in 2000 (Ha Thang Long, 2000)
Species recorded: *P. cinerea* (sighting and living specimen)

The EPRC received two confiscated Grey-shanked doucs (January and August 1997; Nadler, 2000) which were poached in Ba To District.

During a FZS survey conducted in June 2000, more detailed information was collected. A group of 7 to 10 individuals was seen in the Goi Mo mountains. Following records from hunters and locals it is estimated that about 8 groups (Ba To, Goi Keng mountain, Goi Ru mountain, Goi Mo mountain) with a total of 100 to 140 animals live in this area.

An Lao District (BINH DINH)
Special use forest: None
Douc status: Occurence confirmed, last evidence in 2001 (Nadler, pers. comm.)
Species recorded: *P. cinerea* (interview and living specimen)

The EPRC received four confiscated Grey-shanked douc langurs (April 1996, August 1998, December 2000, February 2001; Nadler, 2000; Nadler in prep.) which were poached in An Lao District.

During a FZS survey conducted in May and June 2000 more detailed information of was collected. The species is well known in the area but also heavily hunted. The doucs are obviously easy to hunt in the very poor secondary forest patches. Some groups are already isolated due to agriculture. According to reports of locals and hunters there are about 7 groups with 70 to 85 individuals remaining in the area.

Mom Ray National Park (KON TUM)
Special use forest: National park
Douc status: Provisional occurence, last report in 1995 (Do Tuoc, 1995)
Species recorded: *P. nigripes* (specimen); *Pygathrix* sp. (specimen)

Several museum specimens belonging to Black-shanked doucs are reported from the Mom Ray area. One adult male skull (ZMVNU 867) was collected in May 1980 by Tran Hong Viet in the Dak Wang locality (north of the park). One adult male skull (ZMVNU 851), two juvenile male skulls (ZMVNU 853 and 854) and two adult female skulls (ZMVNU 855 and 868) were collected in May 1980 and December 1981 by Tran Hong Viet in the R'Khoi (=Ro Koi) Commune (northeast of the park). One female skin (ZMVNU 725) and three adult male skulls (ZMVNU 869, 870 and 877) were collected between April 1980 and January 1983 by Tran Hong Viet in Sa Son Commune (east of the park).

Du Tuoc (1995) identified specimens belonging to both *nigripes* and *nemaeus* in the national park.

The taxonomic determination of the doucs living in Mom Ray area is complicated by the discovery of several specimens that do not show all the characteristics of any known species. One juvenile male skin (ZMVNU 726), collected in June 1980 in Sa Son Commune (east of the park) is most likely *P. nigripes*, but has whitish wrists. Several other specimens, though showing mainly the features of *P. nemaeus*, have blackish shanks. This is the case for one adult female (ZMVNU 849), one adult male (ZMVNU 850) collected in January 1978 by Hiet in Mo Ray Commune (south of the park) and one adult of unknown sex (ZMVNU 732) collected in May 1980 by Tran Hong Viet in the Dak Su locality (north of the park). Furthermore, a living animal observed in the guest house of Kon Tum town by Vu Ngoc Thanh and Lippold, which was reported from Mom Ray, also seems to be an intergrade form (Vu Ngoc Thanh, pers. comm.).

The evidence of *P. cinerea* occuring in the reserve, reported by Lippold & Vu Ngoc Thanh (1999), is not considered as reliable here. This record is based on specimens observed in the houses of wildlife traders of Plei Ku City (Gia Lai Province), which are reported to have been caught in the Mom Ray area. However, Plei Ku is a trade post to the west of the Central Highlands. Several specimens of other species seen by A. Tordoff (pers. comm.) in this locality had come from as far as Cambodia. Therefore, we should not draw any conclusions from such specimens.

Kon Plong District (KON TUM)
Special use forest: None
Douc status: Occurrence confirmed, last evidence in 2001 (Eames *et al.*, 2001)
Species recorded: *P. cinerea* (specimen), *Pygathrix* sp. (sighting)

Two to three individuals attributed to *P. nigripes* were seen in February 2000 by an FFI team during an elephant survey conducted in the district (Trinh Viet Cuong, 2000). The sighting was made between 900 to 1,000m a.s.l. in a rich forest near Nuoc Ma stream of village No. 1 (Ngoc Tem Commune) (14°45'21.0"N / 107°22'48.7"E).

A WWF- Indochina Programme survey conducted in December 2000-January 2001 observed a group of 5-6 animals, including two juveniles. The sighting was made at an altitude of approximately 1,400m. a.s.l. in ridge-top primary forest. The team also observed another group of about five individuals in logged evergreen forest near the Nuoc Ca River. Unfortunately, the dim light and speed with which the group ran away prevented a conclusive identification. However, a skin found in a local house in Kon Plong town was identified as *P. cinerea* (Eames *et al.*, 2001). The owner of the skin stated that the animal had been caught in the Mount Ngoc Boc area. It is likely that the two groups observed previously were also of this taxon, given the identification of the skin.

During a FZS survey conducted in June 2000 a first population census was made. According to information from locals and hunters, eight groups occur in this area with 85 to 120 Grey-shanked doucs. This species is well known by the locals and is heavily hunted. Two juvenile doucs were caught in April and May 2000 and sold to an animal trader. The hunters reported that the population had decreased very fast, but there is no knowledge of laws concerning the protection status.

Kon Ha Nung area (GIA LAI)

Special use forest: Partly included in Kon Cha Rang and Kon Ka Kinh Natures Reserves
Douc status: Occurrence confirmed, last evidence in 1998 (Vu Ngoc Thanh, pers. comm.)
Species recorded: *P. nemaeus* (sighting and specimen in Kon Ka Kinh, specimen in Kon Cha Rang), *P. cinerea* (specimen between the two nature reserves); *Pygathrix* sp. (sighting and specimen in Kong Cha Rang)

Le Xuan Canh (1995a) conducted a survey for large carnivores in Kon Ha Nung area in October 1994. The author reported sightings of douc langurs but did not provide details.

Douc langur sightings in Kon Cha Rang and Kon Ka Kinh Nature Reserves are reported by Lippold and Vu Ngoc Thanh from four field surveys conducted in 1994, 1995, 1996 and 1997 (Vu Ngoc Thanh, pers. comm.). Sightings of groups with 51 individuals were reported by Lippold (1998). According to this survey, it appears that *P. nemaeus* and *P. nigripes* are at least parapatric in the area. In Kon Cha Rang, although only the Black-shanked douc was observed on several occasions in the forest, specimens of both the Black- and the Red-shanked douc were seen in hunters' houses (within the forest) and in local markets. However, Lippold & Vu Ngoc Thanh (1995) also report Hatinh and Phayre's langurs from this locality, which is extremely dubious given current knowledge about the ranges, distribution and habitat requirements of both species. Furthermore a photograph labelled as *Pygathrix nemaeus nigripes* in Lippold (1998), is actually a Grey-shanked douc (Nadler, 1997a). Consequently, doubt is cast upon Lippold's records of Black-shanked douc langurs in Kon Cha Rang Nature Reserve (which may have been misidentifications of Grey-shanked douc langurs), and further surveys are required to clarify the taxon's status at the locality.

In Kon Ka Kinh, further to the south, animals observed in the forest (several sightings) and specimens seen in the villages belonged only to the Red-shanked form. (Vu Ngoc Thanh, pers. comm.). No additional confirmation could be made of the sympatry of Black- and Red-shanked taxa in the area as claimed by Lippold (1995a).

Between February and April 1999, a survey was conducted by BirdLife and FIPI for the investment plan of Kon Ka Kinh Nature Reserve (Le Trong Trai, 2000) and, in April 1999, a feasibility study was conducted by FIPI to establish Kon Cha Rang Nature Reserve (Le Trong Trai, 1999b). No animals were recorded in the forest, however, two Grey-shanked douc specimens were seen in a taxidermist's shop in K'Bang town. These animals were reported to have been shot in Dak Roong Forest Enterprise, between Kon Ka Kinh and Kon Cha Rang Nature Reserves.

According to Le Trong Trai (pers. comm. 2000), doucs seem to be rare in the area.

A Yun Pa Nature Reserve (GIA LAI)
Special use forest: Nature reserve
Douc status: Provisional occurrence, last report in 2000 (Tran Quang Ngoc *et al.*, 2001)
Species Recorded: *Pygathrix* sp. (specimen)

The lower jaw of a Black-shanked douc langur (the identification remains questionable) was seen in the area by a BirdLife and FIPI team in April 2000 (Tran Quang Ngoc *et al.*, 2001). The species, locally named "con giac", was reported as uncommon. The area is dominated by a dipterocarp forest habitat.

Ea Kar District (DAK LAK)
Special use forest: Partly included in Chu Hoa Nature Reserve
Douc status: Provisional occurrence, last report in 1996 (Le Xuan Canh *et al.*, 1997a)
Species recorded: *Pygathrix* sp. (interview and specimen)

In April and May 1997, during a survey conducted by IUCN and WWF for large mammals, two photographs of one captive animal in Buon Ba village were examined. According to the authors, the specimens fitted neither the black nor red forms. (Le Xuan Canh *et al.*, 1997a). However, the succinct description given suggests that it was *P. nigripes*.

Yok Don National Park (DAK LAK)
Special use forest: National park
Douc langur status: Provisional occurrence, last report 1994 (Ha Thang Long & Le Thien Duc, 2001)
Species recorded: *P. nigripes* (museum specimen)

One specimen of a young female Black-shanked douc langur was seen in the collection of Tay Nguyen University during a FZS survey in December 2001. According to the inventory of the collection the animal was bought about 1994 in Buon Me Thuot and hunted in Yok Don National Park.

Ea Sup District (DAK LAK)
Special use forest: None
Douc status: Provisional occurrence, last report in 1996 (Le Xuan Canh, 1997a)
Species recorded: *P. nigripes* (interview)

According to the forest staff of Ye Lop Forest Enterprise, Ea Sup District, interviewed during a large mammal survey carried out by WWF and IUCN in June 1997, douc langurs may occur in Tieu Teo mountain (northeast of the district) (Le Xuan Canh, 1997a). According to Vu Ngoc Thanh (pers. comm., 2000), the form recognised was *P. nigripes*. The last dated sighting was in late 1996. The species was not recorded during a survey for the green peafowl in the area by BirdLife and IEBR in March 1998 (Brickle *et al.*, 1998).

Cu Jut District (DAK LAK)
Special use forest: None, partly included in proposed extension of Yok Don National Park
Douc status: Occurrence confirmed, last evidence in 1999 (Trinh Viet Cuong Ngo & Ngo Van Tri, 2000)
Species recorded: *P. nigripes* (sighting); *P. cinerea* (interview)

In December 1999, FFI conducted an elephant survey in the district (Trinh Viet Cuong & Ngo Van Tri, 2000). Five doucs were observed on a single occasion near the Dak Klau Stream in secondary evergreen forest (west of the district), less than 200m north of the base camp (12°43'27"N / 107°42'59"E) at 300 m a.s.l. The animals were about 30m from the observer and were positively identified as *P. nigripes*. One group of 6 individuals was also observed by the team at 12°43'48"N / 107°42'51"E. However, the animals showed features of the Grey-shanked species.

During a FZS survey in December 2001 one hunter identified two different forms of douc langurs. He pointed out that usually the Black-shanked douc langur occurs in the area, while the Grey-shanked douc very rare, and he had never seen both species together in one group. The last sightings reported from different locals were:

July 2001-one group with 50-60 individuals in Dak Klau area; August 2001- one group with 15-20 individuals in Ea Mao area (Chief of FPD Dak Wil); September 2001- one group with about 15, another with over 20 individuals in Ea Mao and Dak Drich areas; December 2001- one group with 20-30 individuals in Dak Drich area.

Dak Mil District (Dak Lak)
Special use forest: None
Douc langur status: Provisional occurrence, last report 2001 (Ha Thang Long & Le Thien Duc, 2001)
Species recorded: *P. nigripes* (interview, specimen)

During a FZS survey, locals reported sightings of Black-shanked douc langurs. Last reports: April 2001 one group with 5-7 individuals north-west of Dak Lao commune, one group with about 20 individuals in Dak Lup area.

A hunter in Dak Lao village reported that he had shot more than 100 Black-shanked douc langurs between 1984 and 2001.

Dak R'Lap District (DAK LAK)
Special use forest: None
Douc langur status: Provisional occurrence, last report 2001 (Ha Thang Long & Le Thien Duc, 2001)
Species recorded: *P. nigripes* (interview)

During a FZS survey in December 2001 local people reported that, close to the borderline to Cambodia, two groups of Black-shanked douc langurs occur with 27 to 30 individuals in total. The last sighting was at the end of November 2001 in Quang Truc village. One animal from a group of 10 was shot.

M'Drak District (DAK LAK)
Special use forest: None
Douc status: Occurrence confirmed, last evidence in 1992 (Pham Nhat, 1994a)
Species recorded: *P. nigripes* (sighting and specimen)

One adult male, skin and skull (FCXM 05/3) was collected in June 1992 by Hoang Son. This specimen seems to be like *P. nigripes* but has white wrists (Fooden, 1996). Pham Nhat (1994a) observed one group of 27 individuals in June 1992, which he reported to be *P. nigripes*. Over a five day survey in June 1997, the same author saw no douc langurs (Pham Nhat, pers. comm., 2000). BirdLife and IEBR surveyed this area for the green peafowl (Brickle *et al.*, 1998) in March to May 1998. Doucs were reported by a local guide to be uncommon.

Krong No District (DAK LAK)
Special use forest: Partly included in Nam Nung Nature Reserve
Douc status: Provisional occurrence, last report in 2001 (Ha Thang Long & Le Thien Duc, 2001)
Species recorded: *P. nigripes* (interview, specimen)

During a FZS survey a hunter from Dac So village reported a sighting in Nam Nung mountain in December 2001 of a group of 5-6 indivuduals from which he shot one subadult animal of 4 kg. Remains of the hunted animals were seen.

Da Lat Plateau (DAK LAK, LAM DONG, KHANH HOA and NINH THUAN)
Special use forest: Partly included in Chu Yang Sinh, Bi Dup-Nui Ba and Deo Ngoan Muc Nature Reserves
Douc status: Occurrence confirmed, last evidence in 2002 (Ha Thang Long, 2002)
Species recorded: *P. nigripes* (sighting and specimen);

Several museum specimens come from the area. One adult female (AMNH 69556), one adult male (ZRC 4-547) and one juvenile male (ZRC 4-548) were collected in March 1919 by C. B. Kloss in Don Duong (=Dran). One adult male (AMNH 69555) was collected by C. B. Kloss in April 1919 at 1,500m a.s.l. in Da Lat. One juvenile male (FMNH 46509) was collected by Osgood in February 1937 on Nui Ba (=Nui Lang Bian) at 1,500m a.s.l..

The species was also reported by local hunters on Hon Nga-Cong Troi-Bi Doup massif during a survey conducted by Eames & Robson (1993) in the area in June 1991. Eames & Nguyen Cu (1994) reported that douc langurs were heard on Nui Bi Dup (=Bi Doup) at 2,000m a.s.l. in December 1993 during the feasibility study to establish Thuong Da Nhim Nature Reserve (currently included in Bi Dup-Nui Ba Nature Reserve). In January 1994 the same team observed two individuals then one group of 18 individuals, with at least two females nursing young, along Ea K'tour in Chu Yang Sin Nature Reserve.

During a survey conducted for large carnivores in November 1994, Le Xuan Canh (1995a) reported a sighting of doucs on Bi Dup Massif but provided no additional details.

Lippold and Vu Ngoc Thanh surveyed the area on two occasions. In May 1998, several dead specimens were seen in wildlife trader houses in Da Lat and Duc Trong Cities. One living animal, kept as a pet, was seen in Da Lat City (Vu Ngoc Thanh, pers. comm.).

The Da Lat Plateau is the origin of the type and paratypes of the *P. n. moi* described by Kloss in 1926: two adult males (BMNH) 1908.11.1.2 and 1908.11.1.3) and one adult female (BMNH) 1908.11.1.4) collected by Dr Vassal in June 1908 on Nui Ba (=Nui Lang Bian); one adult male (BMNH) 1947.1501), two adult females (ZRC 4-549 and 4-550), one infant male (ZRC 4-551) and one juvenile male (ZRC 4-552) collected by C. B. Kloss in April 1919 at 1,800m a.s.l. on Nui Ba (=Nui Lang Bian).

A ranger of the Bi Dup-Nui Nature Reserve and locals reported to a FZS survey team in July 2002 (Ha Thang Long, 2002) of recent sightings of Black-shanked douc langurs, occassionally in large groups.

Another group of 40-50 individuals was seen in December 2001 on the Lang Bi An Mountain and aother group of 70-100 individuals in May 2002 in the Cong Trai area.

Ranger of the nature reserve recorded a video tape in Spring 2002 showing a group of 5-6 individuals.

Di Linh Plateau (LAM DONG)
Special use forest: Partly included in Nui Dai Binh Nature Reserve
Douc status: Occurrence confirmed, last evidence in 1999 (Vu Ngoc Thanh, pers. comm.)
Species recorded: *P. nigripes* (sighting and specimen)

In Di Linh (=Djiring), one adult male (BMNH) 1927.12.1.8), two infant males (BMNH) 1927.12.1.9 and 1927.12.1.15), two adult females (BMNH) 1927.12.1.10 and MNHN 1929.443), one infant female skin (BMNH) 1927.12.1.11), one juvenile male (BMNH) 1927.12.1.14), 1 female (museum unknown) and 3 skins unsexed (museum unknown) were collected by J. Delacour and H. P. Lowe in March 1927 at 1,200m a.s.l..

In June 1991, Eames & Robson (1993) observed this species twice. On the first occasion, one group, of 10 to 15 individuals was seen on Nui Pantar at 1,350m a.s.l.. The second time, two groups of about 10 and 20 animals were seen on Deo Nui San at 1,200m a.s.l.. The species was also reported by local hunters on Nui Braian and Nui Ta Tung. In the latter locality, one skin of an animal shot for meat was seen in a village.

Pham Nhat observed one group of 22 individuals in Bao Loc District in December 1991 (Pham Nhat, pers. comm.).

Lippold and Vu Ngoc Thanh surveyed the area on two occasions. In May 1998, several dead specimens were seen in wildlife trader houses in Di Linh City. In November 1999, two groups of 15-20 individuals were observed in the vicinity of Bao Loc (Vu Ngoc Thanh, pers. comm.).

One specimen, killed in Loc Bac Forest Enterprise (Bao Lam District) in 1997 was observed and photographed by Andrew Tordoff (pers. comm., 2000) in March 2000 in a taxidermist's house.

Nha Trang District (KHANH HOA)
Special use forest: None
Douc status: Unknown, last report in 1905 (museum specimen)
Species recorded: *P. nigripes* (specimen)

One adult male (BMNH 1906.11.6.1) was collected in November 1905 by J. J. Vassal and reported from Bali in Nha Trang vicinity. Nothing is known about the status of doucs in Khanh Hoa Province.

Nui Chua Nature Reserve (NINH THUAN)
Special use forest: Nature Reserve
Douc status: Occurrence confirmed , last evidence in 2001 (Ha Thang Long, 2001)
Species recorded: *P. nigripes* (sighting and specimen)

During a survey conducted in April 2000, local informants stated that this sub-population of *P. nigripes* may number as many as 100. The population size could not be confirmed, but a group of 25 individuals was observed on the bank of Da Thau stream. Vu Ngoc Long (2001) concludes that probably the largest sub-population of Black-shanked doucs in Vietnam probably occurs in Nui Chua Nature Reserve.

A FZS survey in August 2001 was conducted for a census of Black-shanked douc. One group of about 45 individuals was observed and another one of 2 individuals. According to the information of hunters, locals and rangers (more than 100 people) about 12 to 15 groups occur in the area with a total of 500 to 700 individuals. The groups consist of about 20 to 70 individuals. Despite this number, locals reported a markable decrease over the last ten years (Ha Thang Long, 2001).

The Ra Glai minority in this area tradionally do not hunt the doucs. With the influence of Kinh people during the last decades, hunting pressure and animal trade has increased. In 1991/92 one animal trader offered 5000 VND (about US$ 0.55) for one Black-shanked douc langur. According to hunters, about 200 to 300 individuals were killed during this time.

The confiscation of illegal guns around the Nature Reserve organized by the Forest Protection authorities since 1995 could help to stabilize the douc langur population (Ha Thang Long, 2001).

The EPRC received a sub-adult male in April 2001 and a young female in December 2001 from this locality. The animals were poached in the Nature Reserve and subsequently confiscated.

Ninh Hai District (NINH THUAN)
Special use forest: None
Douc status: Provisional occurrence, last report in 2001 (Ha Thang Long, 2001)
Species recorded: *P. nigripes* (interview)

During a FZS survey conducted in August 2001, the western part of the district, outside of the Nui Chua Nature Reserve was also investigated. Officers of the Song Trau special forest management reported regularly douc sightings during patrols in the Hao Chu Hi mountain range (11°42'N/109°01'E). A hunter confirmed that he had hunted 5 langurs in this area before 1996.

Ninh Phuoc District (NINH THUAN)
Special use forest: None
Douc status: Provisional occurrence, last report in 2001 (Ha Thang Long, 2001)
Species recorded: *P. nigripes* (interview)

During a FZS survey conducted in August 2001 a hunter confirmed the occurrence of Black-shanked doucs on Nui Ba Com mountain.

Bien Lac-Nui Ong Nature Reserve (BINH THUAN)
Special use forest: Nature reserve
Douc status: Occurrence confirmed, last evidence in 2000 (Ngo Van Tri, pers. comm.)
Species recorded: *P. nigripes* (specimen)

Black-shanked douc langurs were reported by Cao Van Sung *et al.* (1993). This occurrence was not confirmed by FFI, during a short field survey of 11 days in September 1999 to assess the elephant status in the area. However, some local hunters reported that the species can be found in some places within the reserve (Ngo Van Tri, 1999b).

During the FFI elephant monitoring workshop conducted in August 2000, Ngo Van Tri (pers. comm.) saw a fresh male Black-shanked douc langur specimen that had been shot by local hunters.

During a FZS survey conducted in August 2001 locals reported the occurrence of Black-shanked doucs and hunted individuals. According to a FPD report one douc was confiscated in December 2000; one male was hunted in February 2001 in Song Phan mountain and is now kept as a specimen in Lac Ha village; a baby was confiscated in spring 2001 and kept by rangers but subsequently died after a few days (Ha Thang Long, 2001).

Bu Dang District (BINH PHUOC)
Special use forest: None
Douc status: Occurence confirmed, last evidence in 1992 (Pham Nhat, pers. comm.)
Species recorded: *P. nigripes* (sighting)

This locality is reported by Pham Nhat (pers. comm.) on the basis of sightings made in 1992.

Bu Gia Map Nature Reserve (BINH PHUOC)
Special use forest: Nature reserve
Douc status: Occurrence confirmed, last evidence 2001 (Ngo Van Tri, in prep.)
Species recorded: *P. nigripes* (sighting) *P. nemaeus* (unknown)

P. nemaeus is reported in the feasibility study of Bu Gia Map Nature Reserve (Anon., 1994d) but no reference is given and the record is somewhat unreliable. During an elephant survey in January 2001, Ngo Van Tri reports the observation of 3 groups of Black-shanked douc langurs in the northern, evergreen forested area of the Nature Reserve (Ngo Van Tri, in prep.).

FPD Binh Phuoc reported a confiscation of 11 Black-shanked douc langurs (9 adults, 2 subadults) in June 1997 to the FZS survey team in December 2001. The animals most probably were hunted in the nature reserve.

Loc Ninh District (BINH PHUOC)
Special use forest: None
Douc status: Occurrence confirmed, last evidence in 2001 (Ha Thang Long & Le Thien Duc, 2001)
Species recorded: *P. nigripes* (sighting, specimen)

During a FZS survey conducted in November/December 2001 an observation was made of a group of 5-6 individuals in Bao Dai-Bao Coc area (12°01' 804 N/106°53' 770 E).

The last sightings reported by locals were: April 2001 - 2 individuals close to Thiet Ke River; June 2001 - 2-3 individuals in Bao Coc area; a group of about 15 individuals near Ca-ra-di highland in November; and 20 individuals in December near the same place.

One animal (about 10 kg) was trapped in Bao Nhom in August 2001. In November 2001 a local bought a young one (700-800g) from a hunter. The animal survived only two days. Parts of a Black-shanked douc langur (head, 2 hands, 2 feet, tail and scrotum) were found in a restaurant in Bu Dop in a bottle with brandy. According to the owner of the restaurant, the douc langur was hunted in Bu Dop area in December 2001.

Locals reported a rapid decline in numbers of Black-shanked douc langurs, due to high hunting pressure.

Cat Tien National Park (DONG NAI, BINH PHUOC and LAM DONG)
Special use forest: National park
Douc status: Occurrence confirmed, last evidence in 2003 (G. Polet, pers.comm.)
Species recorded: *P. nigripes* (sighting)

In 2001 more than 20 observations were recorded, but mostly of small groups of 2-7 animals. Since 2000, there have only been 4 observations of 10-16 individuals in one group.

Frequently, Black-shanked doucs are observed on the Ba Sau trail (G. Polet, pers. comm.).

Dinh Quan District (DONG NAI)
Special use forest: None
Douc status: Provisional occurrence, last report in 1999 (Ngo Van Tri & Day, 1999)
Species recorded: *P. nigripes* (interview and specimen)

One specimen of unknown sex was collected in Tan Phu (=Dinh Quan) and reported by Van Peenen *et al.* (1969). It is conserved in the USNM (Fooden, 1996).

From June to August 1999, FFI conducted a survey to examine the biological diversity of mammals in the area of the Tan Phu State Enterprise (Ngo Van Tri & Day, 1999). Interviews report the occurrence of Black-shanked douc langur. According to local hunters, one group of 7 to 11 individuals was seen in April 1999.

Locations where douc langurs are believed to be extinct

Nui Ba Den Cultural and Historical Site (TAY NINH)
Special use forest: Cultural and historical site
Douc status: Unknown, last report in 1874 (Morice, 1875)
Species recorded: *P. nigripes* (specimen)

One young male (museum unknown) was collected between 1872 and 1874 by A. Morice (1875) and reported from Nui Ba Den (Nui-Ba-Dinh). The small remnant of forest in the area is of poor quality and it is unlikely that doucs still remain (Ngo Van Tri, pers. comm., 2000).

Thonh Nhat and Bien Hoa Districts (DONG NAI)
Special use forest: None
Douc status: Unknown, last report in 1965 (Groves, 1970)
Species recorded: *P. nigripes* (specimen)

One adult male (BMNH 1928.2.24.1) and one adult male skin (ZRC 4.728) were collected in Thong Nhat (=Trang Bom) in June 1918 by C.B. Kloss. One adult male (USNM 356976) was collected in October 1965 (unknown collector) in the same locality. One specimen collected in Bien Hoa (before 1971, collector and museum unknown) is reported by Groves (1970). According to Ngo Van Tri (pers. comm.), no suitable habitat remains in the area.

Discussed records

Pu Huong Nature Reserve (NGHE AN) 19°15'-19°29'N/104°43'-105°00'E
P. nemaeus is reported without reference in the investment plan of Pu Huong Nature Reserve (Anon., 1995b). The locality would be in the most northern part of this species range. It was not recorded in 1995 during the field survey of SEE (Kemp & Dilger, 1996). The douc occurrence in this location needs confirmation.

Nam Nung Nature Reserve (DAK LAK) 12°12'-12°20'N/107°44'-107°53'E
P. nemaeus is reported in the investment plan of Nam Nung Nature Reserve (Krong No District, Dak Lak Province) (Anon., 1994b). However, no reference is given. Therefore, this record is not considered reliable.

Buon Ma Thuot (DAK LAK) ca. 12°45'N/108°05'E

Six adults and two infant Black-shanked douc langurs were collected in Buon Ma Thuot in March 1937 by W. H. Osgood (FMNH 46510 to 46518). However, Buon Ma Thuot is an important trading locality in the area and no precision is given about the exact origin of the specimens.

4.4   Status

All three douc langur species were once widely distributed in Vietnam and common throughout their range. After several decades of intense human pressure, populations have been reduced and fragmented.

Hunting is currently the major threat. Douc langurs have probably undergone a more significant decline than other diurnal primate species in Vietnam, partly as a result of behavioural characteristics which make them easy to hunt. They remain motionless but they are still easily seen by humans (Timmins *et al.*, 1999; Pham Nhat *et al.*, 2000). Douc langur meat is not considered tasty by Kinh people (Eames & Robson, 1993), but minorities in the south appreciate the meat for food (Ha Thang Long & Le Thien Duc, 2001). Most animals are hunted for traditional medicine and sometimes also for sale as pets, even if they do not have much chance of long survival in captivity.

One of the main contributing factors to the rapid decline of the douc langur population in Vietnam is the removal of its natural habitat. A large portion of forest in the centre and south of the country suffered from the massive wartime damage. Post-war human demographic explosion and extensive logging for coffee, rubber and cashu plantation reduced the natural habitat.

The core population of Red-shanked douc langur now resides in Lao PDR. These langurs are still common in some areas of Lao PDR, in particular in the eastern part of the country, south of 18°45'N. However, they are absent or very rare northwest of Nam Kading NBCA (Bolikhamxai Province) and further to the south and west of Xe Pian NBCA (Champaxac Province). Nakai-Nam Theun NBCA and the adjacent Hin Namno NBCA together may possess the largest population of Red-shanked doucs remaining in the world (Timmins & Duckworth, 1999). Doucs have been eradicated from disturbed and fragmented forest. They are particularly susceptible to exploitation in the wildlife trade. Most animals are probably traded in Vietnam and in Thailand. With regard to the status of the taxon in Vietnam, the conservation of Red-shanked douc langurs in Lao PDR is a global priority (Duckworth *et al.*, 1999).

From Lao PDR the distribution area extends along the Annamite Mountain Chain to central Vietnam. This remaining natural habitat is now confined on a narrow band near the Lao PDR border. Most forest at lowland levels has been cleared and no forest remains undisturbed. The eastern slope of the Annamite Chain is still subject to human pressure. Between 1985 and 1998, in some locations, such as Anh Son and Con Cuong Districts (Nghe An Province), and Huong Son and Huong Khe Districts (Ha Tinh Province), the quality of the habitat has been dramatically reduced.

Even in areas which maintain a relatively large and undisturbed habitat, such as Phong Nha or Vu Quang, the populations of douc langur seems to be very small in relation to the same habitat in Lao PDR. For instance, in Phong Nha-Ke Bang National Park, only one small group of douc langur was seen, despite four months of intensive surveying by FFI in 1998.

The Black-shanked douc langur has a fragmented distribution area in south Vietnam. Many subpopulations are under pressure due to forest conversion into plantations and agricultural land and hunting. High hunting pressure along the border with Camdodia originates in army stations (Ha Thang Long & Le Thien Duc, 2001).

Obviously there are only two protected areas with higher and, in some measure, stable populations: Cat Tien National Park (Dong Nai, Binh Phuoc and Lam Dong Provinces) and Nui Chua Nature Reserve (Ninh Thuan Province). In Nui Chua Nature Reserve probably the largest sub-population of this species occurs, with estimated 500 to 700 individuals. The distribution and density in Cambodia is still poorly known.

The Grey-shanked douc langur is listed as one of the world's 25 TOP endangered primate species (Conservation International & IUCN Primate Specialist Group, 2000). However, the distribution and a census of the newly discovered species is still under investigation. The species is also under threat through heavy hunting and habitat destruction. The occurrence in protected areas is only confirmed for Song Thanh Nature Reserve (Quang Nam Province).

The global conservation status of the Red- and Black-shanked douc langurs in the *IUCN Red List of Threatened Species* (Hilton-Taylor, 2000) is **"Endangered"**: **EN** A1cd. The 1997 newly described Grey-shanked douc langur is listed as **"Data Deficient"**.
The three douc langur species should be separately listed. The following status' are proposed:
Red-shanked douc langur: **"Endangered" EN** A1 cd
Black-shanked douc langur: **"Endangered" EN** A1 cd, A2 cd, B2 d
Grey-shanked douc langur: **"Critically Endangered" CR** A2 cd

Pygathrix sp. is classified as **"At Risk"** in Lao PDR (Duckworth *et al.*, 1999).
Red- and Black-shanked douc langur are listed as **"Endangered"** the Grey-shanked douc langur as **"Data Deficient"** in Vietnam's national list (Pham Nhat *et al.*, 1998).

4.5   Recommendations for conservation in Vietnam

The three douc langur species are threatened with extinction in Vietnam. Several measures should be undertaken urgently, both to protect the remaining populations which survive outside protected areas in which they occur and to reinforce their conservation in the Vietnamese protected area network.

4.5.1   Conduct further field status surveys

Further surveys are urgently needed to clarify the current distribution and population status of doucs, mainly in the Central Highlands. The identification and immediate protection of a suitably large forest area is a priority.

Red-shanked douc langur

 *   Douc groups may still occur in remaining forest patches of Quang Tri province. The province has never been the subject of biological survey work, excepted in Dakrong District (Le Trong Trai & Richardson, 1999a). Douc langur status should be assessed.

Grey-shanked douc langur

 *   An extensive survey in Quang Nam, Gia Lai, Kon Tum, Quang Ngai, Binh Dinh and Phu Yen provinces should be undertaken, because few data exists about the douc langur status in these provinces. Although this vast region may support some of the largest forest areas in Vietnam, it remains understudied and under-represented in the Vietnamese protected area network.

Black-shanked douc langur

  * A similar situation is found in Da Lat and Di Linh Plateau. This is an immense mountainous area still largely covered by forest, and it is likely that the region contains one of the largest douc langur populations in Vietnam. However, very few surveys have been conducted and much of the natural forest is outside of the protected area system.

4.5.2. Expansion of the protected area network

  * A major area of natural forest lies in the 60 km between Pu Mat to Vu Quang Nature Reserve. A feasibility study should be undertaken to link these regions with continuous coverage, for conservation purposes. Such a corridor would further increase the viability of these protected areas for the long-term. Moreover, this region is adjacent to Lao PDR with Na Kai-Nam Theun NBCA, which extends to the south to Hin Namno NBCA, where douc langurs are still common.

  * The forest between Vu Quang National Park and Phong Nha-Ke Bang National Park is partly included in Nui Giang Man Nature Reserve. However, the area includes the lands of northern Minh Hoa District (Quang Binh Province), which are agricultural, and excludes natural forest extending to the north near Vu Quang National Park. Surveys are needed to revise the boundaries of the protected area and to ascertain whether a linkage between Phong Nha-Ke Bang and Vu Quang is realistic.

  * Southern Quang Binh Province is still largely forested northwards to Phong Nha-Ke Bang National Park. The extension of the park would assist in the conservation of douc langur. Further studies are necessary to assess the feasibility of such a proposal. According to Le Trong Trai & Richardson (1999a), a linkage of southern Quang Binh forests to Da Krong-Phong Dien Proposed Nature Reserves (Quang Tri and Thua Thien-Hue Provinces) would not be feasible since these two areas are separated by agricultural land, grassland and scrubland.

  * An expanse of lowland evergreen forest currently remains in northern Phong Dien District and southern Da Krong District. The presence of doucs was recently confirmed here (Le Trong Trai & Richardson, 1999a). These areas are recommended for upgrading to "Special-Use Forest" status (Le Trong Trai & Richardson, 1999a).

  * Although they were especially affected by chemical defoliation during the American war, the forest patches extending from Da Krong-Phong Dien to Bach Ma National Park via A Luoi District (Thua Thien-Hue Province) are still large. They could be preserved to rebuild corridors between Bach Ma and Da Krong-Phong Dien. (Le Trong Trai & Richardson, 1999a)

  * The conservation potential of the Son Thanh Nature Reserve (southern Giang and western Phuoc Son Districts, Quang Nam Province) was assessed by WWF and FIPI in 1997 (Wikramanayake *et al.*, 1997). The report also noted the biodiversity value of northern Hien District, close to the border of Thua Thien-Hue Province (in particular the occurrence of a new muntjak species *Muntiacus truongsonensis*). It is also suggested that a transboundary conservation landscape could be created in this area linking Bach Ma National Park to Xe Sa NBCA (Lao PDR), where the Red-shanked douc is reported. A southern extension of this protected area complex along the Lao PDR border has the potential to link Song Thanh Nature Reserve

*   The establishment of Ngoc Linh (Quang Nam) Nature Reserve is in progress. This protected area is adjacent to Song Thanh Nature Reserve and Ngoc Linh (Kon Tum) Nature Reserve. Once established, these three protected areas will form one of the largest areas of continuous conservation coverage in Vietnam (nearly 160,000 ha). The FIPI and BirdLife report about Ngoc Linh (Quang Nam) recommends upgrading this vast protection zone to National Park level (Tordoff *et al.*, 2000a). Wege *et al.* (1999) considers the amalgam of these protected areas to be one of the main priorities for conservation of terrestrial forest in Vietnam.

*   Doucs are still common in Da Lat and Di Linh Plateau. However, these areas are under-represented in the "Special-Use Forest" network and hunting and wildlife trade in this area is particularly intensive. The northeastern part of the region is already covered by two protected areas: Chu Yang Sin (Dak Lak Province) and Bi Dup-Nui Ba (Lam Dong Province) Nature Reserves. An extension to the south of Chu Yang Sin will link it to Bi Dup-Nui Ba. Bi Dup-Nui Ba Nature Reserve should be expanded to the southwest to include the largest extent of natural forest in the area. This continuous area of conservation would cover nearly 100,000 ha. Wege *et al.* (1999) considers this proposed protected area complex to be a priority for conservation of terrestrial forest in Vietnam. Two other areas are included on the 2010 list (list included in the EU-funded project *Expanding the Protected Areas Network in Vietnam for the 21ˢᵗ Century*) and would expand the conservation coverage of the Da Lat-Di Linh Plateau: Ta Dung and extension (17,000 ha), South West Lam Dong Province (27,700 ha) proposed protected areas (Wege *et al.*, 1999).

4.5.3   Reinforcement of conservation in protected area network

Phong Nha-Ke Bang National Park and adjacent Ke Bang Forest in Minh Hoa District is one of the highest priority areas for the conservation of doucs and other primates in Vietnam. Besides the occurrence of two loris species and four macaque species, it is known to be inhabited by two threatened langur and one gibbon species: the Hatinh langur (*T. l. hatinhensis*), the Red-shanked douc langur (*P. nemaeus*) and the Southern white-cheeked crested gibbon (*Nomascus leucogenys siki*). Doucs are rare in Phong Nha-Ke Bang National Park. However, the area may have the potential to support a large population of this species (Pham Nhat & Nguyen Xuan Dang, 1999) in conjunction with adjacent Hin Namno NBCA (Lao PDR) where the species is still common.

Although the extension of the Phong Nha Nature Reserve to include the Ke Bang area and upgrading to National Park status has been established by governmental decree as Phong Nha-Ke Bang National Park, no conservation measures have yet been implemented in the Ke Bang area. Infrastructure development, boundary demarcation, assignment and training of rangers, as well as the development of a National Park Management Plan should be implemented as soon as possible, as the hunting pressure is increasing.

5.0. Snub-nosed Monkeys

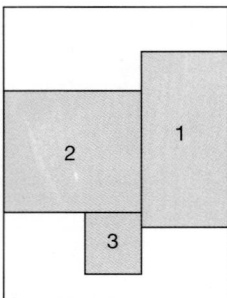

1. Tonkin snub-nosed monkey; ad. male. *Le Khac Quyet*
2. Tonkin snub-nosed monkey; infant about one and a half year old. *T. Nadler*
3. Tonkin snub-nosed monkey; infant. *R. Ratajszczak*

5. Snub-nosed monkeys

5.0 An introduction to the snub-nosed monkeys (genus: *Rhinopithecus*)

5.0.1 Taxonomy

The south-east Asian colobine genera: *Pygathrix*, *Rhinopithecus*, *Nasalis* and *Simias* form the "Odd-Nosed Group" and possess many unusual characteristics of the integument and a high intermembral index. But whether this is more than an informal group is unclear (Groves, 2001).

The relationship between *Rhinopithecus* and *Pygathrix* is disputed. Indeed, the two genera share some characteristics such as the presence of small flaps of skin on the upper margins of the nostrils, the reduction of the nasal bones and a similarity in limb proportions (Groves, 1970; Oates *et al.*, 1994). Following these observations, several authors considered *Rhinopithecus* as a sub-genus of *Pygathrix* (Groves, 1970; Brandon-Jones, 1984; P. Napier, 1985; Corbet & Hill, 1992; Davies & Oates, 1994; Rowe, 1996), but Jablonski & Peng (1993) analysed *Pygathrix* and *Rhinopithecus* and argued strongly for their recognition as a full genus which is now widely accepted (Eudey, 1996/1997; Hilton-Taylor, 2000; Groves, 2001; see also 2.3).

The genus *Rhinopithecus* is commonly named "snub-nosed monkeys". The genus comprises four species, three of them occur in the mountains of southern China, and one, the Tonkin snub-nosed monkey (*Rhinopithecus avunculus*) is endemic to Vietnam.

5.0.2 Morphology, ecology and behaviour

Among "Odd-Nosed Monkeys", *Rhinopithecus* has a distinctive nasal anatomy. The lateral crura of the greater alar cartilages is prominent, the nasal bones are abbreviated or absent and the bony nasal septum and septal cartilage are abbreviated in the rostrum. These features are associated with complex motifs of the mid-facial bony structures. In addition, the interorbital distance is relatively wider and the muzzle more shortened than most colobines, exhibiting a net concave mid-facial profile (Jablonski, 1998). The purpose of this anatomical feature is unclear but it's possible that this could have the function of warming and the humidifying of inhaled air in a dry and cold mountain habitat.

The species of *Rhinopithecus*, in particular the Chinese species, are also distinguished by a series of gnathic and dental charcteristics (Jablonski, 1998).

The Tonkin snub-nosed monkey is Vietnam's largest primate. Compared with other *Rhinopithecus* taxa, *R. avunculus* has a different constitution and different habits (Ren *et al.*, 1998). For instance, *R. avunculus* is more slender, with elongated digits, and slightly less sexual dimorphism.

The diet of the *Rhinopithecus* species seems to consist largely of leaves, grass, lichens, bamboo shots, buds, and fruits. But there is a differentiation within the species (Bleisch *et al.*, 1998).

The diet of *R. avunculus* is composed mainly of fruit and leaves. The first studies of stomach contents suggested that the species was mostly folivorous, with a strong preference for bamboo. Indeed more than 60% of the stomach contents of two females contained bamboo fragments (Ratajszczak *et al.*, 1992). However, the later surveys tend to suggest a more frugivorous diet (Pham Nhat, 1994b; Boonratana & Le Xuan Canh, 1994). This second hypothesis is supported by further studies of stomach contents and field surveys. Boonratana & Le Xuan Canh (1994), using scan sampling from 34 feeding observations, estimated that 47% of its diet constituted whole, unripe fruit, 15% seeds of ripe fruits and 38% of young leaves and leaf parts. No animal remains were found in any of the stomachs examined (Pham Nhat, 1994b). One should be cautious of a determination of diet from these data as some food sources are seasonal and so diet may vary with food availability.

All four species of *Rhinopithecus* have a basic social unit of one-male / multifemale groups and in all four species, these family groups associate in large bands of over 100 animals (in undisturbed populations).

According to Boonratana & Le Xuan Canh (1994), *R. avunculus* form groups with an average group size of 14.8 individuals. Other males form multi-male associations. No solitary animals were ever observed. This social structure is flexible and two or more units often travel, sleep and eat together in larger bands. This arrangement appears to be associated with large overlapping ranges among groups (Boonratana & Le Xuan Canh, 1994). Ratajszczak *et al.* (1992) conducted interviews with local people, which suggested that these large bands could reach more than 100 individuals. However, Ratajszczak *et al.* (1992) considered the latter as the basic unit, which scatter when disturbed.

The Tonkin snub-nosed monkey is diurnal, arboreal, moves quadrapedally and by semi-brachiation. It sleeps in the lower branches of trees and during the period of cold northeast winds, sleeping sites are found close to steep mountain sides (Boonratana & Le Xuan Canh, 1994).

From distant observation, *R. avunculus* is generally silent (Ratajszczak *et al.*,1992). A distinct vocalisation, sounding like a hiccup, may be heard if animals are alarmed or in general communication within or between groups (Boonratana & Le Xuan Canh, 1998; R. Boonratana, pers. comm., 1998).

According to local hunters, after an animal is shot, the group generally does not move immediately but rather remains motionless and silent. This behaviour makes these monkeys particularly vulnerable to hunters, who can kill several animals at one time (Ratajszczak *et al.*, 1992).

The sexual behaviour of *R. avunculus* is unknown. The extreme sexual dimorphism in body size and canine dental complex for some species of snub-nosed monkeys has been interpreted by Jablonski and Pan (1995) as a mechanism by which males compete for mates. That *R. avunculus* is the least sexually dimorphic snub-nosed monkey could indicate that inter-male competition is less important in this species.

5.0.3 Distribution

The zoogeographic range of each *Rhinopithecus* species is very restricted. *R. roxellana* (Golden snub-nosed monkey) is the most widely distributed, occurring in a series of disjunct forests around the periphery of the Sichuan Basin (Sichuan, Gansu, Hubei and Shaanxi Provinces). *R. brelichi* (Guizhou snub-nosed monkey) is restricted to a single population on Mount Fanjing (Guizhou Province). *R. bieti* (Yunnan snub-nosed monkey) is located in five counties in Yunnan Province and Tibet.

All three Chinese species inhabit high mountain forest up to about 4,500m a.s.l. but may descend to lower elevation in winter. Parts of their range are covered by snow for more than half a year.

The Tonkin snub-nosed monkey is endemic to northern Vietnam. Its range is historically limited to areas to the east of the Red River. *R. avunculus* lives at relatively low elevations in subtropical monsoon forests.

5.1 Tonkin snub-nosed monkey
Rhinopithecus avunculus Dollman,1912

5.1.1 Taxonomy

The Tonkin snub-nosed monkey was first described by Dollman (1912) as *Rhinopithecus avunculus* based on eight specimens collected in September 1911 by Owston and Orii in Yen Bai Province (21°42'N / 104°53'E). Subsequently, Pocock (1924) observed several differences from other *Rhinopithecus* species, and placed the species in a new genus *Presbytiscus*. Thomas (1928), who observed 12 specimens of *avunculus*, collected by J. Delacour and H. P. Lowe between 1926 and 1927, supported this hypothesis. He based his conclusion on the structure of the hands and feet, which had longer and more slender fingers. The separation as subgenus was also proposed by Jablonski & Peng (1993), but Davies & Oates (1994) did not see a necessary subdivision and recent taxonomic revisions recognized the Tonkin snub-nosed monkey as a *Rhinopithecus* species (Eudey, 1996/1997; Hilton-Taylor, 2000; Groves, 2001).

5.1.2 Description

Upper parts are dark brown, forehead and cheeks are creamy; the side of the neck is orange buff. The belly and the inner side of the limbs are creamy white. On the outside of the arms and legs a stripe of the same color as the back runs to hands and feet; there is a white patch on the elbows inside the black stripe; there is a buffy white patch on the rump on either side of the tail; there is small orange collar. Face is bluish white with enlarged pink lips.

The tail is very long, with brown to black hairs and a white tassel. The ears are tufted; the hands and feet are black. The nose is upturned and has a tip.

External measurements (adult)

	n	mm	Source
Head/Body length			
male	?	650	Boonratana & Le Xuan Canh, 1994
female	?	540	
Sex?		510-620	Nowak, 1999
Tail length			
male	?	850	Boonratana & Le Xuan Canh, 1994
female	?	650	
Sex?		660-920	Nowak, 1999
Weight		kg	
male	?	14,0	Boonratana & Le Xuan Canh, 1994
female	?	8,5	

DISTRIBUTION OF TONKIN SNUB-NOSED MONKEY (*Rhinopithecus avunculus*) IN VIETNAM
BẢN ĐỒ PHÂN BỐ LOÀI VOỌC MŨI HẾCH Ở VIỆT NAM

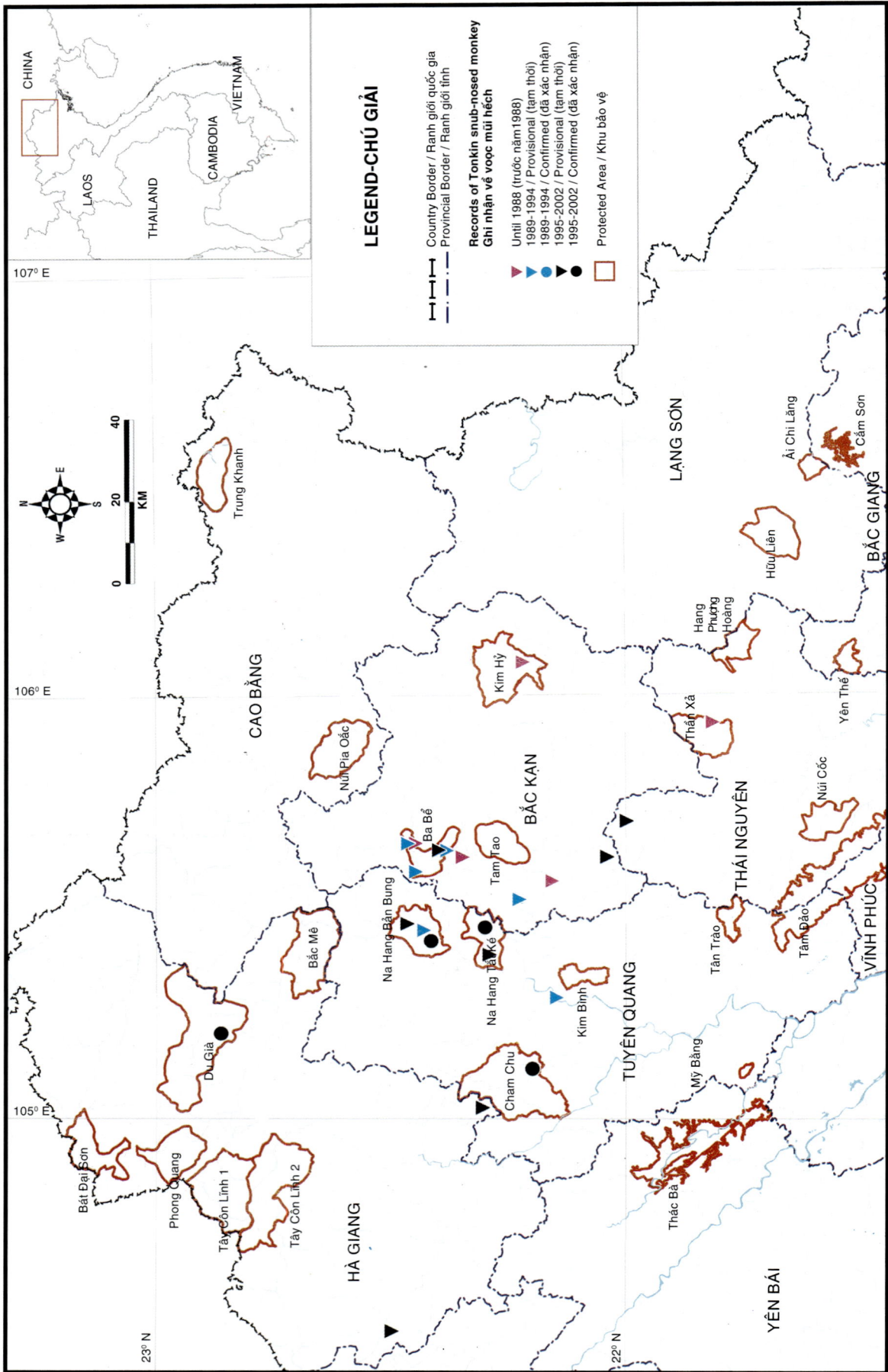

5.1.3 Distribution

The Tonkin snub-nosed monkey is endemic to northern Vietnam. Its range is historically limited to areas to the east of the Red River.

Following the massive deforestation and the intensive hunting of recent decades, its distribution has become dramatically restricted. It is currently only known from small forest patches in Tuyen Quang and Bac Kan Provinces and to a lesser extent in Ha Giang and Thai Nguyen Provinces.

The Tonkin snub-nosed monkey was known from only a few museum specimens, until a WWF team (Ratajszczak *et al.,* 1990) surveyed the area around Ba Be National Park. Ratajszczak *et al.* (1992) conducted the first survey with a special emphasis on *R. avunculus.* Since then, several studies have been conducted, primarily in Na Hang region, Tuyen Quang Province, which have improved our knowledge of the species (Le Xuan Canh *et al.,* 1992; Pham Nhat, 1994b; Boonratana & Le Xuan Canh, 1994; Boonratana, 1998, Dang Ngoc Can & Nguyen Truong Son, 1999; Long & Le Khac Quyet, 2001).

In 2002 a FFI survey team discovered a further Tonkin snub-nosed monkey population in Ha Giang Province (Le Khac Quyet, 2002).

5.1.4 Tonkin snub-nosed monkey records

Duc Xuan Commune, Bac Quang District (HA GIANG)
Special use forest: None
Tonkin snub-nosed monkey status: Provisional occurrence, last report in 1999 (interview) (Dang Ngoc Can & Nguyen Truong Son, 1999)

Chiem Hoa District (Tuyen Quang Province) was visited by a FFI team in November 1999 (Dang Ngoc Can & Nguyen Truong Son, 1999) with a special emphasis on *R. avunculus.* According to three local informants interviewed in Trung Ha Commune (Chiem Hoa District), one group of 8 to 10 individuals still inhabits Duc Xuan forest (Bac Quang District, Ha Giang Province) near the border of their commune.

Na Chi Commune, Xin Man District (HA GIANG)
Special use forest: None
Tonkin snub-nosed monkey status: Provisional occurrence, last report 2001 (interview, specimen) (Le Khac Quyet, 2001)

Le Khac Quyet provisionally confirmed the presence of Tonkin snub-nosed monkeys during a FFI survey (December 2000-January 2001). Interview data suggested 2 groups of 10-20 animals and two tails were found in a local house in Na Chi village.

This very recent data is of enormous significance since it was formerly believed that the Tonkin snub-nosed monkey was confined to only two areas of forest surrounding Na Hang. These new data are in accordance with the species range indicated by Fooden (1996) and Dang Ngoc Can & Nguyen Truong Son (1999) and requires further investigation.

Du Gia Nature Reserve (HA GIANG)
Special use forest: Nature reserve
Tonkin snub-nosed monkey status: Occurrence confirmed, last evidence in 2002 (sighting) (Le Khac Quyet, 2002)

During a FFI survey in January 2002 a population of Tonkin snub-nosed monkeys was discovered in the nature reserve. Locals reported about two groups of Tonkin snub-nosed monkeys with 20 to 30 individuals each.

On 16 January 2002 two local hunters from Khuon Lang village, Tung Ba commune killed one male weighing 16 kg. On three occasions Tonkin snub-nosed monkeys were observed: on 20 January 8 adult animals were seen; the group was probably a male group and estimated to comprise 15 to 20 individuals; on 21 January 13 animals were seen (7 adults, 3 sub-adults, 3 babies); the group was estimated to comprise 25 to 30 individuals; on 22 January probably the same group was seen and 9 adults and 5 babies observed. A local guide counted 27 individuals and the group was estimated to comprise more than 30 individuals.

The whole population is estimated to comprise 50 individuals. This is currently the northernmost population of the species and most probably a relic population which had in the past a larger distribution including Phong Quang and Tay Con Linh Nature Reserves.

Cham Chu Nature Reserve (TUYEN QUANG)
Special use forest: Nature reserve
Tonkin snub-nosed monkey status: Occurrence confirmed, last evidence in 2001 (sighting) (Long & Le Khac Quyet, 2001)

Cham Chu is a mountain range stretching from southeast to northwest and forms the boundary between Chiem Hoa and Ham Yen Districts (Tuyen Quang Province). It is covered mainly with secondary forest, which can be found only above 800m a.s.l. (Dang Ngoc Can & Nguyen Truong Son, 1999).

Several snub-nosed monkey specimens have been collected in Chiem Hoa District. One male skin (IEBR 1437) was collected in the district (date and collector unknown). One adult male (ZMVNU 139) and two adult male skulls (ZMVNU 146 and 544) were collected in November 1962 (unknown collector). In December 1962, an adult female (ZMVNU 140) was collected in Phuc Thinh locality, about 10 km east of Cham Chu (collector unknown) (Fooden, 1996).

In March 1992, Ratajszczak *et al.* (1992) reported the occurrence of *R. avunculus* on Cham Chu mountain on the basis of interviews. Local hunters reported 20 to 40 individuals. Furthermore, two skulls, a dried hand and a foetus were seen in a H'mong house on the slope of the mountain. These specimens were the remains of two animals shot from a group of 15 animals two weeks previously. One of the skulls is now kept in the Forestry College of Xuan May (FCXM 010).

In November 1999, FFI conducted a survey in the district with a special emphasis on the Tonkin snub-nosed monkey (Dang Ngoc Can & Nguyen Truong Son, 1999). The survey was primarily concentrated on the northeastern part of Cham Chu mountain, included in Trung Ha and Ha Lang Commune territories.

Interviews conducted in Trung Ha Commune report that snub-nosed monkeys still occur on the mountain range. A group of 22 to 25 animals were reported by 10 interviews on Khau Vuong and Khau Cang mountains (22°14'N/105°04'E). In this area, a Tay hunter met a group of 30 animals in 1990 and killed an adult female, while 4 other individuals may have been killed in 1994. In 1998, a Tay hunter from Khuon Pong village shot 3 animals. To confirm his claim, he showed to the team the three skeletons and the three tail tips. Bones were kept for making monkey balm. The last animals collected by a local hunter in this area, a female and its offspring, were reported to have been shot in September 1999.

Three other informants claimed that a group with 8 to 10 members still lives on Khuon Nhoa mountain (22°15'30"N / 105°02'15"E). The last sighting was in 1998.

Interviews conducted in Ha Lang commune are concordant. One group of 25 to 30 animals is reported to be often seen in the Khuoi Muoi area (22°12' 40"N / 105°06'10"E), close to Cham Chu peak. One H'mong hunter may have shot one female there in 1998. The snub-nosed monkey's range also may cover Khau Sang mountain. Local hunters shot two individuals in this area in 1992, the two skulls of which were collected by Ratajszczak *et al.* (1992). One male may have been killed in 1995 as well. Two rattan cutters from Hiep village were reported to have seen one group of 10 to 12 animals in Khau Sang mountain in November 1999.

During another FFI survey conducted in May to June 2001 (Long & Le Khac Quyet, 2001) a group of eight Tonkin snub-nosed monkeys was observed in primary forest belonging to Phu Luu Commune.

Following sightings and reports by locals, Tonkin snub-nosed monkey group may, therefore, exist in these areas (Long & Le Khac Quyet, 2001):

* 8 males in one group: Quan Tien, Goc Dinh, Bo De, Su Say, southern Khau Vuong

* 20 animals: Khau, Su Say, Quan Tien, Bo De,

* 20 animals: Khau Sang, U Tum, Coc Nghe, Be Cham Chu, Tham Chang

* 22 animals: Dan Khao, Cot Moc, Man Phu, Dan Puc, Dan Tao

The estimated population for the greater Cham Chu area is 70 animals. Therefore, the Cham Chu population represents one of the two most critical sites for the conservation of *R. avunculus* in the world.

Kien Thiet Forest, Yen Son District (TUYEN QUANG)
Special use forest: None
Tonkin snub-nosed monkey status: Provisional occurence, last report in 1992 (interview) (Ratajszczak *et al.*, 1992)

In March 1992, Ratajszczak *et al.* (1992) reported that local hunters estimated 20 to 40 snub-nosed monkeys still living in the area. However, this information could not be confirmed. No further reports are available.

Na Hang Nature Reserve (TUYEN QUANG), and vicinity in Cho Don District (BAC KAN)
Special use forest: Nature reserve (except for contiguous forest area in adjacent Cho Don district)
Tonkin snub-nosed monkey status: Occurrence confirmed, last evidence in 2000 (sighting, vocalization record) (Martin, 2000)

The first record of this area is of one adult male skin (IEBR 782) collected in Thanh Tuong Commune in October 1965 by Ngo Van Phu. An adult female skull (FCXM 008) collected by FIPI is reported from Na Hang District (date unknown) (Fooden, 1996).

During a survey of primates in north Vietnam (Ratajszczak *et al.*, 1990), snub-nosed monkeys were reported in Ban Thi Commune (Cho Don District, Bac Kan Province) which forms the boundary with the Ban Bung sector of the nature reserve.

A survey for snub-nosed monkey was conducted in Tat Ke and Ban Bung sectors in March 1992 (Ratajszczak *et al.*, 1992). In Tat Ke sector, based on interviews, the surveyers estimated that 40 to 50 individuals in one group lived on Loung Nioung Mountain. They are not known to live permanently on Pac Ta Mountain. The authors report that at least 11 animals were killed by H'mong and Dao hunters during the three first months of 1992. One fresh skull of an old animal was observed by the team.

Based on interviews with local hunters they estimated that 90 to 110 animals live in the Ban Bung sector. The occurrence was confirmed by a brief sighting of one group of 12 individuals. This was the first sighting of the Tonkin snub-nosed monkey in the wild. In addition, the team saw two freshly killed female adults (one female skull is conserved in FCXM, No 007; museum unknown for the second skull). One of the females had been nursing a female infant, which was caught alive, but later died in Hanoi Zoo, killed by a bandicoot rat. The stomach contents of the three animals were examined (Ratajszczak et al., 1992; Pham Nhat, 1994b).

Another sighting was obtained in December 1992, by Le Xuan Canh et al. (1993). The team briefly observed 20 to 30 animals in Tat Ke sector. In addition, vocalisations were heard in January 1993.

Several sightings were reported during a survey conducted between September 1993 and February 1994 (Boonratana & Le Xuan Canh, 1994). In Tat Ke sector, the presence of three harems and two all-male groups was confirmed. 80 individuals were estimated, of which at least 72 were observed (number given during a single count). In Ban Bung sector, two harems were observed. The authors estimated 50 individuals, 23 of which were observed. This estimation was made from two observations in two different places on the same day.

SEE conducted two successive surveys, one in Ban Bung sector from January to March 1996 (Hill et al., 1996a), followed by another one in Tat Ke sector from July to September 1996 (Hill & Hallam, 1997). The team failed to observe any animals in Tat Ke sector. The report noted that, food supplies being abundant in the wet season, the groups may fragment to breed, which explains this failure. The team only observed snub-nosed monkeys twice in Ban Bung sector including a single sighting of a group of four individuals.

Snub-nosed monkeys were heard in March 1998 by Tan Kit Sun (Singapore Zoo) and B. Martin (ZSCSP) in Tat Ke sector (B. Martin, pers. comm., 2000).

In July 1998, FFI, with financial support from IUCN Netherlands, started a programme to strengthen the management and protection of the reserve (Boonratana, 1999). In Tat Ke sector, three sightings and two vocalization recordings of snub-nosed monkeys were made. At least one harem with 13 individuals was observed with one adult male and at least three adult females. One sighting of 35 animals was reported in the Ban Bung area in September 1998 by Mr Luong (IEBR) and Mr Hai (patrolling group). Interviews reported sightings in August 1998 of one group with 20 to 40 animals on a trail between Chiem Hoa District and Bac Kan Province and of one other group with 40 to 50 animals on a trail between Ban Bung and Ban Vi Village (Bac Kan Province).

The last sightings of snub-nosed monkeys in the boundaries of Na Hang Nature Reserve were reported by the patrolling group (tuan rung) of the Tonkin Snub-nosed Conservation Project (Martin, 2000) in Tat Ke sector; 16 individuals (10 adults, 6 youngs) were observed in August 1998 by Mr. Thong (ranger) in the Thom Bac area. From October 1999 to April 2000, Mr. Tinh (ranger) observed the snub-nosed on five occasions. In the Khau Tep area, he saw one group of over 30 animals (with one juvenile) in September and October 1999. A short film sequence was made during the first sighting. In the Thom Bac area, he observed one group of over 10 animals in October 1999, over 15 in November 1999 and over 18 in April 2000, most probably the same group.

In October-November 1999 FFI carried out a survey to assess the status of primates in Tuyen Quang and Bac Kan Provinces (Dang Ngoc Can & Nguyen Truong Son, 1999). Three H'mong hunters reported one group with 40 to 45 individuals on the border between Xuan Lac (Cho Don District, Bac Kan Province) and Vinh Yen Communes (Na Hang Nature Reserve) (22°18'30"N / 105°28'00"E). One animal, weighing 12kg was reported to have been shot in this area in 1995. Snub-nosed monkeys were occasionally seen in 1997 and 1998. Sightings in 1995 of a group of 18 to 20 individuals were also reported on the boundary between Xuan Lac and Thanh Tuong Communes (Na Hang Nature Reserve). Three interviews reported that an animal was killed in this area in 1992. In the north of Ban Thi

Commune (Cho Don District, Bac Kan Province), close to the border with Thanh Tuong Commune (Na Hang Nature Reserve), the last sighting was reported in 1994-1995 by two H'mong hunters. A survey carried out by FFI/ PARC in January 2003 obtained further confirmtion from hunters that Tonkin snub-nosed monkeys have not been seen in the forest between Ban Thi and Xuan Lac communes (Cho Don district) since 5 years. A hunter from Da Na village (Xuan Lac) shot 2 individuals in 1987 from a group of 20. Tonkin snub-nosed monkeys might still occasionally range into the Ban Thi-Xuan Lac area as the forest is contiguous with Na Hang Nature Reserve. One local hunter reported from second hand information that in 2002/3 Tonkin snub-nosed monkeys occasionally came across from Na Hang Nature Reserve (Le Khac Quyet & Trinh Dinh Hoang, 2003a). As habitat extension this forest is important and considering its habitat integrity and biological diversity the area should be designated as species and habitat conservation area (Le Trong Trai et al., 2001; Momberg & Fredrickson, 2003).

According to the different sightings reported in the reserve, snub-nosed monkeys are now very rare. In Tat Ke, it is not clear if there are two distinct groups (one over 30 and one over 10 animals) or if one frequently splits up. In Ban Bung sector, only one group of 35 individuals is confirmed. From the experience of Mr. Tinh (ranger) it seems like the home range of the Tonkin snub-nosed monkey group in the Tat Ke sector does not exceed 9 km^2 (Martin, 2000). It seems that one group of a relatively large size may occur on the boundary of Ban Bung sector and Xuan Lac Commune (Bac Kan Province). The latter locality is outside the boundary of Na Hang. With regard to the extremely low density of animals in Na Hang, which contains the largest confirmed population in Vietnam, it is a priority to extend the boundaries to the east.

It is not possible to estimate the current number of Tonkin snub-nosed monkey in Na Hang. However, it is unlikely that the population exceeds 100 individuals in the whole reserve (R. Boonratana, pers. comm., 2000).

Ba Be National Park and proposed extension (BAC KAN)
Special use forest: Partly included in Ba Be National Park
Tonkin snub-nosed monkey status: Provisional occurrence, last report in 1997 (interview) (Nong The Dien, pers. comm., 2000)

According to a Ratajszczak et al. (1990) survey in 1989, no informants could confirm the occurrence of snub-nosed monkeys in Ba Be National Park. However, a first SEE survey in the park from July to September 1994 (Kemp et al., 1994) reported from interviews of local hunters and park staff that the species still inhabited Ba Be forest. The report of the second survey conducted in October-December 1996 concluded that it was unlikely that any animals survive there (Hill et al., 1997). According to local hunters, animals were seen on Dau Dang Mountain (close to Nang River) in 1997, though it was the last record in this area (Nong The Dien, pers. comm., 2000).

It is thought possible that some groups still occur in the proposed extension to Ba Be National Park. Indeed, local hunters report that three groups live on Phia Booc mountain, Dong Phuc Commune, in the extreme southwestern part of the proposed extension. One skull was collected in this area (Nong The Dien, pers. comm., 2000).

Tat Pet area, Cho Don District (BAC KAN)
Special use forest: None
Tonkin snub-nosed monkey status: Provisional occurrence, last report in 1999 (interview, specimen) (Dang Ngoc Can & Nguyen Truong Son, 1999)

Tat Pet is the border area of the communes of Phong Huan, Yen My and Dai Sao in Cho Don District (Bac Kan Province). The natural habitat has been reduced by intensive exploitation of bamboo. The remaining forest is now located in the northeastern part of Phong Huan Commune and the eastern part of Yen My and Dai Sao Communes. The primary forest is largely confined to steep hills.

In 1989, during a primate survey in north Vietnam (Ratajszczak et al., 1990), snub-nosed monkeys were reported by local people in the vicinity of Phong Huan Commune.

In October and November 1999, FFI carried out a short survey to assess the primate status in this area. 12 hunters interviewed in different villages reported one group of about 6 individuals. In 1998, one female (7.5 kg) was shot in Tat Pet (22°05'03"N / 105°36'51"E, 390m) by a Tay hunter of Phong Huan Commune. One left hand, kept for medicinal purposes, was photographed by the team.

Khuoi Muoc and Khuoi Chang areas, Cho Don District (BAC KAN) and Dinh Hoa District (THAI NGUYEN)
Special use forest: None
Tonkin snub-nosed monkey status: Provisional occurrence, last report in 1999 (interview, specimen) (Dang Ngoc Can & Nguyen Truong Son, 1999)

Khuoi Muoc and Khuoi Chang are two mountains making the border between Yen Nhuan and Yen My Communes (Cho Don District, Bac Kan Province) and Linh Thong Commune (Dinh Hoa District, Thai Nguyen Province). The forest is very fragmented and not able to support any significant population of primates.

One adult male skull (ZMVNU 787) was collected in Linh Thong by Hoang Dinh Mac in August 1966. One adult male skull (ZMVNU 789) was collected in the same locality by Luu Dinh Kieu in March 1967.

In October and November 1999, FFI carried out a short survey in these communes to assess the primate status. According to local hunters, snub-nosed monkeys still occur in this area.

Three hunters reported one group of 18 to 20 individuals in Khuoi Muoc (22°01'40"N/105°39'50"E). In 1998, two Tay hunters encountered this group and shot one individual. To confirm this information, one right hand and one left foot were shown to the FFI team. In this area, the last sighting was reported in April 1999 (Dang Ngoc Can & Nguyen Truong Son, 1999).

According to nine local informants, another group of 4 individuals may inhabit the Khuoi Chang area near Khuoi Phuong (22°01'40"N/105°39'04"E). In 1997, one animal was killed from a group of 7 animals in this area. In October 1998, another hunter shot an adult male from a group of 6 animals. Its skeleton was photographed by the team. On the same mountain, another adult male was killed. Hands and feet kept as medicine were observed. The last sighting was reported in October 1999 (Dang Ngoc Can & Nguyen Truong Son, 1999).

Ngoc Phai Commune, Cho Don District (BAC KAN)
Special use forest: None
Tonkin Snub-nosed monkey status: Provisional occurence, last report in 1993 (museum specimen, FCXM)

This locality is reported by Ratajszczak et al. (1990) from their survey in 1989. Local informants claimed that a small group of snub-nosed monkeys was occasionally seen on the upper slopes of Lung Luong Mountain. In addition, a skull was observed in a hunter's house at the base of this mountain. One adult male skull (FCXM 009) was collected in September 1993 by Nong The Khiem in Ngoc Phai.

Yen Tu Nature Reserve (QUANG NINH, BAC GIANG)
Special use forest: Nature reserve
Tonkin Snub-nosed monkey status: Provisional occurrence, last report 2002 (Le Khac Quyet, 2002)

FFI surveyed the area in 2002. Interview data and observed dead specimens suggest that 5-7 individuals might still be present (Le Khac Quyet, 2002).

Recent unpublished reports

In 2001 FFI started a Tonkin snub-nosed monkey conservation project. At the time of printing this report three surveys had been conducted with this species as the primary focus. All information is provisional but presented here to highlight the need for such surveys in areas with promising interview data.

In Ha Giang, interview data suggests that the species is still present in small numbers in Xin Man, Bac Quang and Quan Ba Districts (Luong Van Hao. pers, comm.; Le Trong Dat, pers. comm.). No information on the range of groups was obtained, so at this stage it is impossible to say how many groups may survive and whether they are connected to each other and those in the Cham Chu area.

Locations where Tonkin snub-nosed monkeys are believed to be extinct

Tay Con Linh I Nature Reserve (HA GIANG) 22°51'N / 104°49'E

R. avunculus is reported in the investment plan for Tay Con Linh I Nature Reserve (Anon., 1994e). However, no references are given, and furthermore, a recent FFI survey conducted in December 2000-January 2001 suggests the species is now extinct in this locality based on interview data (Le Khac Quyet, 2001).

Luc Yen District (YEN BAI) ca. 22°05' / 104°43'E

In May 1984 one adult male (skin conserved in FCXM, not numbered) and one adult female killed by local hunters were seen by Pham Nhat (Ratajsczcak *et al.*, 1992). The female was nursing an infant, which was kept as a pet and died after six days (Pham Nhat, 1994b). The stomach contents of the three animals were analyzed (Ratajsczcak *et al.*, 1992; Pham Nhat, 1994b). Pham Nhat (1991) who visited again this place in 1990, found that this area had been totally deforested following the building of the dam on the Red River.

North of Ra Ban Commune, Cho Don District (BAC KAN) ca. 22°11'N / 105°39'E

In 1989, during a survey for primates in north Vietnam (Ratajsczcak *et al.*, 1990), snub-nosed monkeys were reported by local people in the hills north of Ra Ban Commune. However, when FFI visited the district in January 2000, this information could not be confirmed (Phung Van Khoa & Lormee, 2000). Three hunters of Ra Ban were interviewed and did not recognize pictures of the species.

Kim Hy Nature Reserve (BAC KAN) ca. 22°14'N / 105°59'E

Geissmann and Vu Ngoc Thanh (1998) reported that a Tay hunter living close to the Kim Hy locality claimed that snub-nosed monkeys still occur in Kim Hy forest. Despite this, the authors noted that it was probably extirpated from this area. Some Kim Hy hunters had a name ("*Ca Dac*") for this species and reported that it had been driven away by gold mining activities (A. Tordoff, pers. comm., 2000).

Than Xa Nature Reserve (THAI NGUYEN) ca.21°49'N / 105°56'E

The species is believed to have occurred there in earlier years (Geissmann & Vu Ngoc Thanh, 2000).

Discussed records

Phong Quang Nature Reserve (HA GIANG) 22°57'N / 104°56'E

R. avunculus is reported in the investment plan for Phong Quang Nature Reserve (Anon., 1997b). However, no reference is given. It was reported in interviews carried out during a recent FFI survey, but the threat of landmines near the Chinese border prevented investigation in the forest (Le Trong Dat, pers. comm.).

Tam Tao Nature Reserve (BAC KAN) 22°18'N / 105°35'E

In the official letter No. 102/CV-UB from Cho Don District People's Committee proposing the Tam Tao Nature Reserve, *R. avunculus* is quoted as present. Surveys in this area conducted by FFI/PARC in January 2003 could not confirm the presence of the species. Hunters from Na Sam village reported that Tonkin snub-nosed monkeys were present until 1975 and are now locally extinct. The habitat has been seriously degraded, and the gazettement of a nature reserve is unjustified (Le Khac Quyet & Trinh Dinh Hoang, 2003).

Dai Tu District (THAI NGUYEN) ca. 21°37'N / 105°38'E

R. avunculus is reported in Dai Tu District in the *Red Data Book of Vietnam* (Ministry of Science, Technology and Environment, 1992). However, no references are given.

Hoanh Bo District (QUANG NINH) ca. 21°10'N / 107°00'E

One skull was reported to have been collected in Quang La Commune. Ratajsczczak *et al.* (1992) considered the occurrence of snub-nosed monkeys in this locality as questionable. Their report noted that it was likely that the skull had been kept by hunters in order to sell and that the animal was not caught in the locality where it was collected. Furthermore, BirdLife and FIPI, who visited the district in November 1999, affirmed that no informants were familiar with this species (Tordoff *et al.*, 2000b).

Bac Kan (BAC KAN), Tuyen Quang (TUYEN QUANG) and Yen Bai (YEN BAI)

Specimens were collected in Bac Kan (=Bach Thong) (Napier, 1985; ZMVNU 147), in Tuyen Quang (Lapin *et al.*, 1965) and Yen Bai (Napier, 1985). However, these locations present the main cities of the provinces and, therefore, we cannot put forward any conclusions about the exact provenance of the specimens.

5.1.5   Status

Currently, there are only three known locations about with recent evidence where Tonkin snub-nosed monkeys occur: Na Hang (2 sub-populations), Cham Chu and Du Gia Nature Reserves. The estimated numbers of the populations are 95-130 animals in Na Hang (two sub-populations), 30-70 animals in Cham Chu, and 21-50 animals in Du Gia.

For five other areas there is information about only very small or single groups:

1. 8-10 animals: Duc Xuan Commune, Bac Quang District (HA GIANG)

2. 10-20 animals: Na Chi Commune, Xin Man District (HA GIANG)

3. 5-7 animals: Yen Tu Nature Reserve (QUANG NINH, BAC GIANG)

4. 18-20 animals: Khuoi Muoc and Khuoi Chang areas, Cho Don District
                    (BAC KAN) and Dinh Hoa District (THAI NGUYEN)

5. 6 animals: Tat Pet Area, Cho Don District (BAC KAN)

The total number of Tonkin snub-nosed monkeys sighted is 123 individuals. The population is estimated at 93 to 307 individuals.

Historically restricted to a small range, the Tonkin snub-nosed monkey is naturally vulnerable. Threats to the species and it's habitat are accelerating due a major dam construction project on the Gam river and several small and medium-scale minining operations in Na Hang, Cho Don and Du Gia.

5.1.5.1   Hunting

The most immediate threat to the survival of *R. avunculus* is hunting. Despite claims from various sources that the species is not hunted, the surveys obtained information to the contrary (Long & Le Khac Quyet, 2001, Le Khac Quyet, 2002).

The species is shot whenever encountered by hunters, even though it is not the prime target of hunting trips. The meat is not regarded as tasty, but the bones are used to make "monkey balm" which can be sold for high prices in the traditional medicine trade. The only reason why more individuals have been not shot was purely assigned to lack of opportunity due to the rarity of the species and not because of law enforcement or awareness of the species' importance (Long & Le Khac Quyet, 2001). Evidence of four animals shot in 2000 and 2001 (Long & Le Khac Quyet, 2001) and one animal in 2002 was obtained by Le Khac Quyet (2002).

Dang Ngoc Can & Nguyen Truong Son (1999) reported to have seen at least 8 shotguns in Phong Huan, 11 in Yen Nhuan, 9 in Ban Thi, 8 in Trung Ha and 7 in Ha Lang. H'mong people commonly make their own shotguns. A compilation of the surveys conducted in 1992 (Ratajszczak *et al.*, 1992), 1999 (Dang Ngoc Can & Nguyen Truong Son, 1999), 2001 (Long & Le Khac Quyet, 2001) and 2002 (Le Khac Quyet, 2002) shows that at least 40 animals have been shot in the last 12 years (Table 5.1.5.1-1).

Table 5.1.5.1-1:
Number of Tonkin snub-nosed monkeys reported to have been killed between 1990 and 2002 in Bac Kan, Thai Nguyen and Tuyen Quang Provinces.

Area	Reported Date	Number of individuals killed	Reported Weight (in kg)	Remains found
Cham Chu Nature Resrve (TUYEN QUANG)	2000/2001	4		
Tat Pet, Phong Huan-Yen My-Dai Sao comm., Cho Don Dist. (BAC KAN)	1998	1	7,5	1 left hand
Khuoi Chang-Khuoi Phuong, Yen Nhuan-Yen My Comm., Cho Don Dist. (BAC KAN) and Linh Thong, Dinh Hoa Dist. (THAI NGUYEN)	1997 1998 1998	1 1 1	9-10 9-10	1 skull 1 right hand and 1 left foot
Khuoi Muoc, Yen My Comm., Cho Don Dist. (BAC KAN) and Linh Thong Comm., Dinh Hoa Dist. (THAI NGUYEN)	1998	1	12	
Border between Xuan Lac Comm., Cho Don Dist. (BAC KAN) and Na Hang Dist. (TUYEN QUANG)	1987 1992 1995	2 1 1	? 10 12	
Khau Vuong, Trung Ha Comm., Chiem Hoa Dist. (TUYEN QUANG)	1990 1994 1994 1998 9/1999	1 3 1 3 2	18 15, ?, ?, 13 6, 7, 12	3 skeletons and 3 tails
Khuoi Muoi, Ha Lang Comm., Chiem Hoa Dist. (THAI NGUYEN)	1992 1998	2 1		
Khau Sang, Ha Lang Comm., Chiem Hoa Dist. (THAI NGUYEN)	1995	1	14	
Tat Ke, Na Hang Dist. (TUYEN QUANG)'	1992	11		
Ban Bung, Na Hang Dist. (TUYEN QUANG)'	1992	3		2 dead females and 1 living infant
Du Gia Nature Reserve (HA GIANG)	2002	1	16	
TOTAL		42		

5.1.5.2   Dam construction

For the populations of Na Hang Nature Reserve, the most serious threats are posed by the Gam river hydropower and flood prevention dam project. Construction began in autumn 2002. Some 8500 workers will move into the area for dam contruction. The current workforce has already reached 2000. It is expected that the current population in Na Hang of 4700 people will raise to 15,000 at the height of the construction boom (EIA 2002). This will lead to increased demand for wildlife products and firewood.

The direct threat to Na Hang Nature Reserve due to inundation is limited to 220 hectares of mostly secondary forest along the banks of the Gam and Pac Vang rivers.

Increased activities in the area due to improved accessibility by roads and future lake, noise and vibrations from blasting in construction and quarry sites, noise from construction and traffic will cause stress to wildlife populations that can translate into decreased breeding patterns, increased suspectibility to predation, and avoidance of disturbed areas (even if they are important feeding areas).

5.1.5.3   Mining

Mining exploitation is a serious threat to habitat integrity due to increased demand for wildlife products, fire wood and construction timber. No environmental impact assessment has been undertaken in Vietnam for small and medium-scale mining operations.

This activity is widespread in limestone areas where the largest remaining snub-nosed monkey population occurs. Mines of zinc and aluminum have been based in Ban Thi Commune (Cho Don District, near the Na Hang Nature Reserve boundary) for several decades. By the year 2000 about 500 m^3 of wood per year for securing mine shafts were extracted from the Ban Thi/Xuan Lac forest and only by June 2002 the mining company started to source the wood from the Cho Don state forest enterprise. In 2002 the 700 workers created an increasing demand for wildlife products and firewood supplied by local villagers. With the completion of a zinc oxidation facility in 2005 the work force is expected to reach 1700.

Gold mining is a major activity in the Tat Ke sector in Na Hang Nature Reserve. When gold is found, an area as large as 100 ha may be cleared for prospecting. Furthermore, mining operations also increase the number of loggers, firewood collectors, and hunters in the region (Ratajszczak *et al.*, 1992; Dang Ngoc Can & Nguyen Truong Son, 1999).

In Minh Son commune, Bac Me district, adjacent to the Tonkin snub-nosed monkey habitat in Du Gia Nature Reserve, the provincial state mining company started in February 2003 the development of a mine for zinc, tin and lead exploitation. The expected work force is 200 employees.

5.1.5.4   Deforestation, habitat degradation and fragmentation

Deforestation, habitat degradation and fragmentation are serious threats in all sites. This is largely due to agricultural encroachment, fire wood extraction and timber exploitation, as well as road contruction leading to habitat fragmentation. In addition, the collection of forest products such as bamboo and rattan, for both local and commercial usage, is common in some areas (Ratajszczak *et al.*, 1992; Dang Ngoc Can & Nguyen Truong Son, 1999).

The major problem for the long term survival of the species is the fragmentation of the population into 9 sub-populations. No interbreeding between sub-populations is possible.

5.1.5.5 · Conservation status

The Tonkin snub-nosed monkey is one of the world's TOP 25 most endangered primate species (Conservation International and IUCN Primate Specialist Group, 2000).

The species is considered as **"Critically Endangered"** in the *IUCN Red List of Threatened Species* (Hilton-Taylor, 2000) with the criteria **CR** C1, E. As result of the current status, *R. avunculus* should be listed with the criteria: **CR** B1, B2 abde, C1. (A quantitative analysis for the criteria E does not exist). In the Vietnamese national list the species also has the status **"Critically Endangered"** (Pham Nhat *et al.*, 1998)

5.1.6   Recommendations for conservation in Vietnam

The conservation of the Tonkin snub-nosed monkey is one of the highest priorities in the domain of environmental protection in Vietnam. Immediate measures must be taken in order to increase the chances of survival for this species, which is almost condemned to extinction.

5.1.6.1   Conduct further field status surveys

* In Ha Giang Province there is only the recent sighting in Du Gia Nature Reserve. This province is within the historical range of this species. Further surveys should be carried out to assess the status of the species in this province

* Two locations, Kinh Thiet Commune (Yen Son District, Tuyen Quang Province) and Ngoc Phai Commune (Cho Don District, Bac Kan Province) were respectively recorded in 1992 and 1993. However, these places have not been visited recently. According to forest cover, it is unlikely that these areas can support a large population of snub-nosed monkeys. However, due to the global critically endangered status of this species, even the smallest groups should be identified, perhaps for a future translocation or captive breeding programme.

* Further surveys are urgently needed in Phia Booc area (southeast of the proposed extension of Ba Be National Park). Information from local people reveals that the species may still occur in the area.

5.1.6.2   Expansion of protected area network and improvement of protected area management

* The two sectors of Na Hang Nature Reserve may possess the largest populations of *R. avunculus*. Special-use forest status has been approved by a Tuyen Quang People Committee's Decision. However, the reserve has not yet been established by governmental decree so there is still the potential to revise the boundary in light of recent findings. The FFI survey in October-November 1999 found that the animals living in Ban Bung may cross the border with Ban Thi and Xuan Lac Communes (Cho Don District). The forested limestone hills located in these areas should be included in a future extension of the reserve (Boonratana, 1999). For administrative reasons and to enhance local community collaboration it may be better to gazette the forest in Xuan Lac commune adjacent to Na Hang Nature Reserve and Ban Thi commune as Species and Habitat Conservation area under the management of the commune and Cho Don District. With the presence of two globally threatened primate species, *T. francoisi* and *R. avunculus*, Na Hang Nature Reserve is of the highest priority for primate conservation.

∗ The second largest known population of snub-nosed monkey may be located in the Cham Chu Nature Reserve (Chiem Hoa and Ham Yen Districts). A long term study with a special emphasis on this species should be conducted. A community-based conservation programme should be initiated as well as a training and capacity building programme for rangers to improve law enforcement.

∗ To secure the population in Du Gia Nature Reserve, a strictly protected core zone through consensus with local stakeholders need to be designated. A local patrol and monitoring group should be recruited from local villagers. As this is the easiest location for sightings of the Tonkin snub-nosed monkey, this site offers ideal conditions for long-term ecological studies.

5.1.6.3   Mitigation of impacts from dam construction

To mitigate potential impacts from the Gam river dam construction, the following measures should be undertaken in collaboration with 'Electricity of Vietnam Company' (EVN), Song Da 9 construction company and relevant government agencies at provincial and national level:

∗ Develop a financing mechanisms for the long-term protection of the Gam river watershed and the Tonkin snub-nosed monkey with EVN.

∗ Employ and train additional rangers to oversee critical construction phases, vegetation clearing, resettlement and operation phases.

∗ Control restaurants and bushmeat trade, conduct spot checks of vehicles moving out of the area.

∗ Conduct regular awareness work among construction workers.

∗ Monitor the provision of coal for cooking by Song Da 9 construction company.

∗ Lobby for resettlement of villages out of Na Hang Nature Reserve.

5.1.6.4 Other options for conservation

Currently, no information is available to conclude with any degree of certainty that a viable population of *R. avunculus* remains in Vietnam. The last hope for the survival of the species may be a captive breeding programme which might be undertaken immediately.

The Biodiversity Action Plan for Vietnam (Government of the Socialist Republic of Vietnam & Global Environment Facility Project, 1994) lists *ex-situ* conservation projects as part of a coordinated species survival strategy. *Ex-situ* programs are recommended for taxa that meet three criteria:

* The taxa has a restricted area, endemic range with a global range of less than 50,000 km²

* The taxa is "Critically Endangered"

* *In-situ* conservation is failing to prevent population decline.

R. avunculus meet all three criteria.

Hence, a timely assessment of the behaviour, diet and ecological relationships of this species is necessary to support both a captive breeding programme and the required planning for subsequent release into a suitably designed, well-protected natural area which can support these species.

If small, isolated groups are found, an alternative to captive breeding would be to translocate animals to areas with larger sub-populations, provided that threats to the species in the release area have been eradicated and sufficent data made available on the viability of the habitat and forest area to sustain additional animals.

6.0. Threats to primates in Vietnam

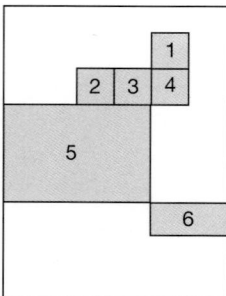

1. Hunters with living macaque. *T. Nadler*
2. Delacour's langur, stuffed infant. *M. Kloeden*
3. Delacour's langur, stuffed specimen for sale. *T. Nadler*
4. Tonkin snub-nosed monkey, stuffed head. *Le Khac Quyet*
5. Delacour's langur heads and hands in brandy as "medicine". *T. Nadler*
6. Rangers confiscate a Grey-shanked douc langur, kept as pet. *T. Nadler*

6. Threats to primates in Vietnam

6.1 Hunting and wildlife trade

Hunting is the main threat to primate populations in Vietnam. This activity has a considerable impact on wildlife in the country due to high human density, easy access to guns and the existence of a well-connected wildlife trade in South-east Asia.

The use of guns to hunt is widespread and largely uncontrolled. Possession of firearms is common and they are easily available. Some weapons were kept by people after the wars, however, homemade guns are also commonly used. In 1992, in a two month period, Nghe An Province policemen checked and confiscated 10,124 weapons including 3,829 army guns, 5,895 shotguns and 317 sport guns (Nhan Dan Newspaper No. 15150, cited by CRES 1997). Shooting animals is often opportunistic and every mammal and bird species constitutes a potential target. Use of non-selective trapping is also common and widely reported.

Although the proportion of wildlife in local diet is not well assessed, it is clear that it represents a protein source, mainly in the remote mountainous areas where agricultural resources are poor (e.g. Che Tao Commune in Yen Bai Province, Geissmann et al., 2000). Local people traditionally eat all primate species. Being a target of relatively large size, primates are systematically shot when encountered. However, if we consider the extremely low density of primates in Vietnamese forests, it cannot be expected that they constitute an essential food source for local human populations.

Several interviews report that the meat of leaf-monkeys is not tasty and so unpopular (Ratajszczak et al., 1990; Nadler & Ha Thang Long, 2000; T. Nadler, pers. comm.; Dang Ngoc Can, pers. comm.). However, reports of local consumption of these animals have been documented and may not be uncommon (Eames & Robson, 1993; Duckworth et al., 1999; Ngo Van Tri & Lormée, 2000).

The main threat is posed by commercially oriented hunting. Wildlife constitutes a large part of the local income through markets. Primates are sold for food, medicine, pets or for various decorative purposes. The internal wildlife trade in Vietnam is particularly active. However, a number of people involved in this lucrative trade point out that most animals are destined to be sold in China. It is clear that China has played, and continues to play, an important role in biodiversity loss in Vietnam. Furthermore, Vietnam is the center of wildlife traffic in Indochina with the trade being well organised (Compton & Le Hai Quang, 1998). Wildlife is bought in Lao PDR (Nooren & Claridge, 2001) or Cambodia (Martin & Phipps, 1996), then brought as far as China to supply its rampant wildlife trade. In addition, since hunting in Vietnam is not profitable due to the rarity of valuable animals, Vietnamese hunters often act in adjacent Laotian forests (T. Osborne, pers. comm., 2000; Duckworth et al., 1999; Timmins & Evans, 1996).

It is probable that the main motivation to hunt primates in Vietnam is for medicine production. This activity is widely reported in several areas. It may constitute the most serious threat for the survival of primate populations. Evidence of commercial exploitation of primates for this purpose suggests that this activity is particularly intensive and is a major concern. During one month of surveying, Dang Ngoc Can & Nguyen Truong Son (1999) found the remains of eight Tonkin snub-nosed monkeys shot during the last ten years which were kept for traditional medicinal purposes. The case of the Tonkin snub-nosed monkey reveals the level of hunting for such a purpose. This species is eaten and is not kept as a pet because it cannot survive in captivity without proper care. However, the population has been seriously reduced throughout its range. In Phong Nha-Ke Bang area, Timmins et al. (1998) cited various sources reporting organised hunting groups from local communities going into the forest to hunt diurnal primates with the apparent objective to sell the animals for medicinal preparation. Each year several tonnes of dried carcasses are prepared in this way for sale (Pham Nhat & Nguyen Xuan Dang, 1999).

Primates are also sold as trophies. A large quantity of wildlife trophies, including stuffed primates, was reported by Eames & Robson (1993) in Da Lat and along the road between Da Lat and Bao Loc cities (Lam Dong Province), principally for sale to local tourists or as ornaments in local hotels. Ghazoul & Le Mong Chan (1994) reported the same activity in Sapa market (Lao Cai Province). There have been decreasing signs of such trade in Sapa in 1997 and 1998, suggesting a reinforcement of controls and reflecting a reduced wildlife density (Tordoff *et al.*, 1999).

The possession of wildlife as pets is widespread in Vietnam. Among primates, macaques are the most commonly encountered. However langurs, gibbons and lorises have also been reported. It is still easy to find live primates in markets, even though they are protected by Vietnamese law.

Zoological gardens attract many visitors in Vietnam and contain a varied set of exotic species. Primates are a very popular and important attraction. Unfortunately, a number of the animals come directly from the wild despite the illegality of such trade.

6.2   Habitat disturbance

Forest destruction in Vietnam was massive during the last half of the 20th century and although is now somewhat controlled it still continues.

Strategic herbicide spraying and intensive bombing by American forces during the American war (1963-1975) contributed to the destruction of large forested areas mainly in the centre and the south of the country. It has been estimated that about 22,000 sq. km of agricultural land and forests were destroyed during this period (Collins *et al.*, 1991). Although the forests of northern Vietnam did not suffer as greatly from the direct effects of the war, indirect impacts were considerable. For instance, bombing of the north of the country resulted in an exodus of people from the plain of the Red River Delta to mountainous areas, where they subsequently cleared forest to create agricultural land (Pham Binh Quyen & Truong Quang Hoc, 1997).

After the war, the demographic explosion and the subsequent increase in demand for agricultural land led to a considerable reduction of the remaining forest cover. This phenomenon was particularly apparent in the lowlands and in the northern part of the country where the human density is higher. Particularly in northern Vietnam, most of the lowland forests were lost and montane forests have been significantly reduced and highly fragmented. Between 1943 and 1995, natural forest cover in Vietnam declined from 44% to 28% (Wege *et al.*, 1999).

Although the central region was previously cleared at a slower rate, this trend may soon be reversed. In the first place, remnant forest blocks in other parts of the country are often too inaccessible to be the subject of profitable commercial logging or to be converted into agricultural land. In the second place, human population density in the Central Highlands is increasing due to the transmigration of people from the over-populated northern provinces. This region received 600,000 migrants between 1976 and 1988 (Pham Binh Quyen & Truong Quang Hoc, 1997). Dak Lak Province, which had the highest immigration rate of any province in Vietnam, saw its human population rise from 1,026,000 to 1,242,000 between 1990 and 1995 (Brickle *et al.*, 1998). Some of these migrants were officially resettled as part of government population redistribution programmes but a large number of spontaneous migrants followed them. The latter were often relatives or friends of officially resettled people.

Besides agriculture encroachment, various disturbances in the natural habitat are caused by the collection of timber and non timber forest products: rattan, bamboo, fuelwood, fruits, honey, *Cinnamomum* sp. tree oil, and medicinal plants.

Timber extraction by state forest enterprises and illegal loggers has resulted in the loss and degradation of large areas of forest. From official sources, it has been estimated that about 80,000 ha of forest were degraded for this purpose in 1991 (Pham Binh Quyen & Truong Quang Hoc, 1997).

Timber or non timber forest product collectors, who must often spend several days in the forest, subsist on natural resources, including primate meat. Trees supporting fruits or bee nests are sometimes cut down to facilitate collection. The collection of *Cinnamomum* sp. tree oil has been reported to have a severe negative impact on the environment. The distillation process involves boiling the wood, which requires a large quantity of firewood. It is estimated that for each tree distilled, 10 ha of forest is negatively affected (Lambert *et al.*, 1994; Le Trong Trai *et al.*, 1996).

Habitat destruction is also associated with hydro-electricity, and other infra-structural development activities. Hydro-electric dams cause a three-fold problem. Firstly, the reservoir may inundate large areas of forest. It has been estimated that as much as 30,000 ha of forest is lost per year due to the creation of reservoirs (World Bank, 1995; cited by Pham Binh Quyen & Truong Quang Hoc, 1997). Secondly, human settlements must be relocated in other places, which often means forest areas. Thirdly, areas surrounding the reservoir become more accessible to loggers and hunters. Such consequences are illustrated in Luc Yen District (Yen Bai Province), where forest has been entirely cleared following the building of the dam on the Red River (Pham Nhat, 1991). Currently several other dam projects are in place in Vietnam, including one operation in Na Hang Nature Reserve (Tuyen Quang Province) one of the last refuges of the critically endangered Tonkin snub-nosed monkey.

Development of the road network in an area of high biodiversity value does not only involve the clearance of large areas of forest during construction. A road often also constitutes an impassable barrier for many species, facilitates the exploitation of forest products, and may act as a focus for settlement of migrants. These problems are associated with the current project to develop the Ho Chi Minh (HCM) national highway, linking Ha Noi to Ho Chi Minh City. This new link will pass through several areas of high importance for primate conservation. Of particular note is a link road between highway 20 and the HCM highway which is currently undergoing construction. This will bisect Phong Nha-Ke Bang National Park (Quang Binh Province). The park possesses the most important population in the world of the endangered Hatinh langur (*Trachypithecus laotum hatinhensis*) as well as 9 other species of threatened primates. This new route will seriously jeopardise a number of these species' sub-populations. Ngoc Linh Nature Reserve (Quang Nam Province) is similarly afflicted by the HCM highway, one of the only places where the poorly known and recently described Grey-shanked douc langur (*Pygathrix cinerea*) has been recorded.

Special mention must be made of mining activities, principally gold mining. These have been frequently reported in limestone areas in northern Vietnam and are of high importance in terms of primate conservation in areas such as Kim Hy, Bac Kan Province (Geissmann & Vu Ngoc Thanh, 2000; Ngo Van Tri & Lormée, 2000), and Na Hang, Tuyen Quang Province (Ratajszczak *et al.*, 1992; Dang Ngoc Can & Nguyen Truong Son, 1999). Large areas of forest are cleared for ore exploitation. Furthermore, this activity attracts a large number of workers, who subsist on forest products or exploit them to supplement their incomes.

Deforestation and habitat fragmentation are a major threat to the long term survival of primate populations. Not only do these processes reduce the area of suitable habitat, and the number of animals, but it also increases the accessibility of forest areas to people. Furthermore, these processes, particularly habitat fragmentation, tend to isolate populations, leading to inbreeding. The long-term consequences of inbreeding may include decreased resistance to disease and higher incidence of infertility. The degree to which primates can disperse across areas of non-forest has not been clearly demonstrated but it is unlikely that arboreal primates will travel significant distances across grassland, scrub or cultivated areas to reach nearby forest patches.

7.0. Conservation Policy

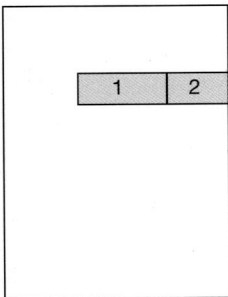

1. Delacour's langur; infant male, four weeks old. *T. Nadler*

2. Primary forest, Cuc Phuong National Park. *T. Nadler*

7. Conservation policy

7.1 Hunting and wildlife trade

Hunting, primarily for commercial purposes, is the principal threat to wild fauna in Vietnam. Primate species particularly suffer from this activity due to their high value in wildlife markets. Urgent policy measures must be undertaken to control hunting and wildlife trade and to effectively apply the current laws.

* Controlling gun possession. Officially a licence is required to keep and use a gun in Vietnam. However, possession of firearms is widespread in the countryside and guns are used without any licence. Several decades of war considerably increased the availability of weapons (Timmins *et al.*, 1998; Nadler, pers. comm.). Some minorities, such as H'mong people in northern Vietnam, also have shotguns, which they traditionally make themselves. In response to the dramatic increase in hunting, gun confiscations have been undertaken in some areas such as Phong Nha-Ke Bang (Le Xuan Canh *et al.*, 1997b; Timmins *et al.*, 1998; Pham Nhat & Nguyen Xuan Dang, 1999) by provincial government authorities. This policy resulted in at least a partial decrease of this form of hunting. Gun confiscation must be continued and generalised, as well as the control of firearm production. However, this activity will be useless and will probably cause more damage to wildlife unless a strict prohibition of snaring and other non-selective trapping is simultaneously undertaken.

* Control of hunting for non-threatened animals. Hunting should not be strictly prohibited but controlled and restricted to certain species of non-threatened animal populations. For this reason, and as stipulated above, only selective hunting must be authorised. If after many years of human exploitation the balance of the ecosystem is disturbed, hunting may be necessary in the future to control the populations which could cause damage to the forest environment. For example, if the ungulate population increases too much due to the extinction of large predators.

* Prohibition of threatened species hunting. According to Vietnamese law (Decree No. 18/ HDBT of January 17, 1992; Decree No. 48/2002/ND-CP of April 22, 2002 and correction 3399/VPCP-NN of June 21,2002) every species of leaf monkey, gibbon and loris is considered under the highest level of protection: group IB. The law (Decree 17/2002/ND-CP of February 8, 2002) strictly prohibits hunting, catching, killing, purchasing, selling, storing, keeping in cages, transporting or using wild animals, including products thereof. For each violation Vietnamese laws (Decree No. 77/CP of November 29, 1996: Decree No. 26-CP of May 7, 1996 and Decree 17/2002/ND-CP of February 8, 2002) enact a fine of 1 to 50 Million VND (=US$ 65-3,250) or criminal procedure. However, it does not seem that the law is effectively applied. For instance, FFI survey teams recorded that at least 13 Tonkin snub-nosed monkeys, 11 Western black-crested gibbons, 3 Eastern black-crested gibbons and 10 Francois' langurs were killed during the last five years. FZS survey teams investigated about 230 Delacour's langurs, about 150 Cat Ba langurs and dozens of douc langurs which were killed since the law was enacted. However, no judicial procedures were enacted against the hunters responsible. Furthermore, most discoveries of animal remains were made with a representative of the local Forest Protection Department, who never appeared concerned by such evidence of illegal activity. Laws must be enforced and corresponding punishments must be effectively applied. Control of any illegal activities must occur and information on the penalties incurred for any law violation must be widely publicised.

* Prohibition of the trade in threatened animals. As well as hunting, trade of protected primate species is strictly forbidden by Vietnamese law and any violation is normally subject to punishment. However, it is still easy to find threatened animals in markets, as much in rural as in urban markets. Furthermore, in several places, wildlife traders act with impunity, though they are known by local authorities. These facts reflect that the law is not properly applied and controls are largely insufficient. A central policy to control commercial use of wildlife must be developed. The prohibition and punishment of production, use and trade of traditional medicine made from primate parts must be a priority for primate conservation given that this is the main motivation for which animals are now hunted. Wildlife trade control must be extended to zoological gardens. Only animals born in captivity or confiscated from illegal trafficing should be authorised to enter such institutions. Although a large proportion of traded animals are probably used locally, a significant part serve to supply neighbouring countries' markets, mainly Chinese (Compton & Le Hai Quang, 1998; Duckworth et al. 1999b, Ngo Van Tri & Day, 1999; Ngo Van Tri & Lormée, 2000). In this way, international cooperation is required, particularly with regard to destination countries, such as China and Thailand. Patrolling groups must be increased along the Chinese border where a significant amount of wildlife products are sold. These measures must be extended along the border of Lao PDR and Cambodia, from whence much of the Vietnamese market is supplied and in turn the Chinese market (Martin & Phipps, 1996; Compton & Le Hai Quang, 1998; Duckworth et al., 1999b).

7.2   Population management

The isolation of forested areas and the reduction of populations due to hunting has resulted in several primate taxa being limited to small, fragmented sub-populations which are unable to interbreed. Recent discoveries of locations where endangered primate species occur, and their dramatic decline in these places suggest that only translocation or ex-situ management could save these individuals and the loss of the genetic diversity.

Single animals and very small populations could be translocated to areas where other groups exist. Unfortunately, at the present time no area in Vietnam possesses the necessary conditions of safety to make such a programme viable.

A captive breeding programme cannot be considered as a priority in relation to the in-situ protection of primate species. However, the Biodiversity Action Plan for Vietnam (Government of the Socialist Republic of Vietnam and Global Environment Facility Project VIE/91/G31 1994) lists ex-situ conservation projects as part of a coordinated species survival strategy. Ex-situ programmes are recommended for taxa that meet three criteria:

* The taxon has a restricted, endemic distribution with a global range of less than 50,000km^2

* The taxon is critically endangered

* In-situ conservation is failing to obstruct population decline.

All three criteria are met for

* Cat Ba langur

* Delacour's langur

* Grey-shanked douc langur and

* Tonkin snub-nosed monkey

The aims of an *ex-situ* programme should be:

* A captive breeding programme to establish a small and stable captive population.

* A release programme to close the gaps between groups in the wild to guarantee genetic flow for the entire wild population, to strenghten declined wild populations or to reintroduce populations in their original habitat where the species was eradicated. The precondition for this step is a safe habitat without poaching and foreseeable habitat destruction.

Confiscated young langurs, and injured or handicapped animals should be remain in captivity. Groups caught for a translocation programme could also be divided, and some members used to establish a captive population. This might be a good way to increase the rate of reproduction and ensure genetic diversity (such as through introducing wild surplus males into a breeding programme).

Zoological gardens attract a large number of visitors each year and primates are among the most popular species. Captive breeding programmes could be conducted in these institutions. This has the double advantage of representing a potential income source and increasing public awareness.

However, due to the extreme precaution that must be taken for such a programme, even if captive management is undertaken in zoological gardens, it must be done in close collaboration with an experienced institution. For many years the EPRC has successfully kept and bred such langur species under captive conditions. The EPRC also has experience managing groups living in semi-wild conditions in two areas surrounded by electrical fencing (Nadler, 1996c, 1997c, 1998b, 1999a, 2000). Thus the EPRC provides an suitable location to start with the *ex-situ* programme.

To summarise, population management in captivity should have a double impact for the conservation of highly threatened species:

* Preservation of the genetic pool of the species

* Value for scientific research and conservation education

7.3   Habitat disturbance

Sustainable income alternatives to destructive forest exploitation should be developed. Given that certain local peoples subsist and are highly dependent on forest resources, no biodiversity conservation programme will be efficient without appropriate alternative income generation strategies for forest product dependent households. This strategy must be economically attractive and must increase the living standard of the rural population. Raintree *et al.* (1999) proposed different strategies including: (1) planting of local forest trees of high timber value, and ordinary species to supply firewood; (2) planting of grafted fruit trees of superior varieties; (3) planting of rattan along field boundaries and in the understoreys of tree crops; (4) planting of medicinal plants in the understoreys of tree crops. These alternatives will be efficient if the local market is improved by business training and organisation of the different steps of production and selling.

The negative impacts of infrastructure development on the environment must be limited. The development of infrastructure such as hydro-electric dams or roads is a necessity for economic progress at both national and local levels. However, often there may be environmental repercussions. Environmental impact assessments should be conducted for each case in order to determine the real economic potential of the project as well as the environmental consequenses. Furthermore, such development projects should contain an in-built management component to prevent the loss of biodiversity and environmental degradation.

7.4   Protected area management

A revision of the current protected area network seems to be necessary. Wege *et al.* (1999) stated that 1,345,000 ha of land were under special-use forest jurisdiction but also stated that 575,000 ha of this area is non-forest land, consisting of scrub, grassland and agricultural land. Several protected area boundaries must be revised to exclude areas which do not support forest. In the same way, other areas of greater biological value should be added to the current protected area network. In terms of primate conservation, Table 7.4-1 lists the proposed new protected areas or extensions, which should allow a better representation of threatened leaf-monkey and gibbon species into the national protected area network.

This status review was restricted to leaf-monkeys. These species were preferred because their global conservation status is of critical concern. However, the status and distribution of lorises and macaque species is also poorly known. It is clear that these two groups are heavily hunted and traded throughout Vietnam. Furthermore, macaques are likely to become highly threatened in the near future due to high hunting pressure and by the increasing use of snares, because these species are more terrestrial (Phong Nha-Ke Bang area, Timmins *et al.*, 1999; Vu Quang National Park, R. Yves pers. comm., 2000).

An urgent need is for the reinforcement of protection measures in the currently protected areas. More than 100 protected areas exist in Vietnam, including nature reserves, national parks and cultural and historical sites. However, for a number of them the current management is insufficient to guarantee any effective protection. In order to improve environmental protection and safeguard important primate populations within the current protected area network in Vietnam, the following activities should be undertaken:

* Intensification of patrols in special use forests

* Reinforcement of guarding protected area borders

* Development of ranger training programmes

Table 7.4-1
Revision of the protected area network recommended for leaf-monkey and gibbon conservation

Proposed new protected area or extension	Province	Target species
Van Ban area (gazettement of a nature reserve contiguous with Hoang Lien Son National Park extending to the border with Yen Bai province)	Lao Cai	*Nomascus concolor* *Trachypithecus crepusculus*
Trung Khanh area (gazettement of a species and habitat conservation area in Phong Nam & Ngoc Khe communes)	Cao Bang	*Nomascus nasutus nasutus*
Xuan Lac area (gazettement of a species and habitat conservation area in Xuan Lac commune, Cho Don district, contiguous with Na Hang Nature Reserve)	Bac Kan	*Rhinopithecus avunculus*
Na Hang Nature Reserve extension (expansion of Tat Ke sector to the North, Duc Xuan & Sinh Long communes, possibly including Lung Nhoi west of the Gam river)	Tuyen Quang	*Trachypithecus francoisi*
Kim Hy area	Bac Kan	*Trachypithecus francoisi,* *(Nomascus nasutus nasutus)*
Cat Ba National Park extension	Hai Phong	*Trachypithecus poliocephalus poliocephalus*
Pu Hoat area	Nghe An	*Trachypithecus crepusculus,* *Nomascus leucogenys*
Khe Net area	Quang Binh	*Trachypithecus laotum hatinhensis,* *Pygathrix nemaeus, Nomascus* *leucogenys siki*
Phong Nha - Ke Bang National Park extension (expansion to include the entire Ke Bang limestone area and Nui Giang Manh or gazettement of new contiguous protected area)	Quang Binh	*Trachypithecus laotum hatinhensis,* *Pygathrix nemaeus, Nomascus* *leucogenys siki*
Ngoc Linh area	Quang Nam	*Pygathrix cinerea, Nomascus* *leucogenys or gabriellae*
Kong Cha Rang-Kon Ka Kinh Nature Reserves extension (link extension)	Gia Lai	*Trachypithecus germaini, Pygathrix* *sp., Nomascus leucogenys or* *gabriellae*
Yok Don National Park extension (extension to the north in Ea Sup District and to the south in Cu Jut District)	Dak Lak	*Trachypithecus germaini, Pygathrix* *nigripes, Nomascus gabriellae*
Proposed protected areas on Da Lat and Di Linh Plateau (see Wege et al., 1999) : Chu Yang Sin extension, Bi Dup-Nui Ba extensions, Ta Dung Nature Reserve, South-west Lam Dong proposed protected area, Kalon Song Mao extension	Lam Dong, Dak Lak, Khanh Hoa and Binh Thuan	*Trachypithecus germaini, Pygathrix* *nigripes, Nomascus gabriellae*

7.5   Education and awareness building

The development of education and awareness building programmes must be sensitive to the specific educational needs of both local- and national-level target audiences. Locally targeted programmes working in communities adjacent to areas identified as priorities for primate conservation must address the needs of adults as well as children. Specific interventions should focus on concepts of extinction and basic ecology as well as seeking to generate and enhance the intrinsic value of living primates.

On a national level there is great potential to generate support for primate protection through media campaigns in press, televison and other popular media such as postage stamps and postcards. Many Vietnamese are not aware of the primate diversity in their country, nor of the threats currently facing primates nationally.

7.6   Research and surveys

The present database is far from exhaustive and must be completed and modified following the results of further surveys. The authors wish for this work to be a benchmark for future national monitoring of threatened leaf-monkeys and gibbons.

Several areas of high potential conservation value for leaf-monkeys and gibbons are both poorly surveyed and under-represented in the current protected area network. Further surveys are recommended in each species chapter. However, resulting from this work, among all the areas it is suggested that further primate surveys will be highly useful in the following provinces for the following target species:

Ha Giang: *Rhinopithecus avunculus, Trachypithecus francoisi, Nomascus* sp.

Kon Tum, Gia Lai: *Pygathrix* sp., *Trachypithecus germaini, Nomascus gabriellae*

Lam Dong: *Pygathrix nigripes, Trachypithecus germaini, Nomascus gabriellae*

8. References

Adler, H. J. 1991. On the situation of the Delacour's langur (*Trachypithecus f. delacouri*) in the north of Vietnam - Proposal for a survey and conservation project. *Primate Report* 31: 6-7.

Adler, H. J. 1992. Cuc Phuong Nationalpark, Nordvietnam, Projekt 1129-91 der Zoologischen Gesellschaft Frankfurt von 1858. *Mitt. Zool. Ges. fuer Arten- und Populationsschutz* (1): 1-6.

Anon. 989. Feasibility Study for Son Tra Nature Reserve, Da Nang City, Quang Nam-Da Nang Province. FIPI-Hanoi. (Vietnamese).

Anon. 1990a. Economic and Technical Report, Krong Trai, Nature Reserve, Son Hoa District, Phu Yen Province. FPD Phu Yen Province, Tuy Hoa. (Vietnamese).

Anon. 1990b. Management Plan for Bach Ma National Park. MS, Hue.

Anon. 1991a. *Rhinopithecus* sp. added to U. S. endangered list. *Asian Primates* 1(3): 4.

Anon. 1992a. Eco-detectives visit Vietnam animal markets. *IPPL News* 192: 3-7.

Anon. 1992b. Meeting on endangered primates in Vietnam. *Asian Primates* 2(3).

Anon. 1993a. Another sighting of *Rhinopithecus avunculus*. *Asian Primates* 2(3-4): 2-3.

Anon. 1993b. Management Plan for Cat Tien National Park. Ministry of Forestry of Vietnam.

Anon. 1993c. Feasibility Study for Hang Kia-Pa Co Nature Reserve, Mai Chau District, Hoa Binh Province. People's Committee of Hoa Binh Province. (Vietnamese).

Anon. 1993d. Project to establish a protected area in Tat Ke Ban Bung, Na Hang, Tuyen Quang. People's Committee of Tuyen Quang Province. (Vietnamese).

Anon. 1994b. Investment Plan for Nam Nung Nature Reserve, Dak Lak Province. People's Committee of Dak Lak Province and MARD. (Vietnamese).

Anon. 1994c. Investment Plan for Du Gia Nature Reserve, Ha Giang Province. People's Committe of Ha Giang Province. (Vietnamese).

Anon. 1994d. Feasibility Study for Bu Gia Map Nature Reserve, Phuoc Long District, Song Be Province. People's Committe of Song Be Province. (Vietnamese).

Anon. 1994e. Investment Plan for Tay Con Linh I Nature Reserve, Ha Giang Province. People's Committee of Ha Giang Province. (Vietnamese).

Anon. 1994f. Feasibility Study for Ba Na Nui Chua, Quang Nam-Da Nang Province. People's Committee of Quang Nam-Da Nang Province and Ministry of Forestry. (Vietnamese).

Anon. 1995a. Feasibility Study and Proposal for Implementation of the Extended Ba Be National Park. People's Committe of Cao Bang Province. (Vietnamese).

Anon. 1995b. Investment Plan for Pu Huong Nature Reserve, Nghe An Province. People's Committee of Nghe An Province. (Vietnamese).

Anon. 1995c. Feasibility Study of the Bi Dup-Nui Ba Nature Reserve in Lam Dong. People's Committee of Lam Dong Province. (Vietnamese, English summary).

Anon. 1995d. The Socio-Economic Situation of Buffer Zone Communities of Bach Ma National Park. Preliminary edition. WWF and European Community (VN 0012.31).

Anon. 1995e. Management Feasibility Study for Expanding Ben En National Park, Nhu Xuan District, Thanh Hoa Province. People's Committe of Thanh Hoa Province. (Vietnamese).

Anon. 1997a. Assessment on Biodiversity and Possibility for the Establishment of a Nature Reserve in the West of Quang Nam Province. WWF, MARD, FPD and FIPI.

Anon. 1997b. Investment Plan for Phong Quang Nature Reserve, Ha Giang Province. Nothwestern Sub-FIPI. (Vietnamese).

Anon. 1997c. Integrative approaches for biodiversity conservation in Vietnam with a case history of the Chu Mom Ray Nature Reserve, Kon Tum Province, Central Highlands. Australian National University, School of Resource, Management and Environmental Sciences.

Anon. 1997d. Proposed Second Revision of the Bach Ma National Park Management Plan. WWF and European Community (VN 0012.01).

Anon. 1998a. Statistic Yearbook of 1997. Government Statistical Office, Hanoi. (Vietnamese).

Anon. 1998b. Investment Plan for Pu Hu Nature Reserve, Thanh Hoa Province. People's Committee of Thanh Hoa Province. (Vietnamese).

Anon. 1999 Feasibility Study to establish Kon Cha Rang Nature Reserve, Gia Lai Province. People's Committee of Gia Lai province, (Vietnamese).

Baker, L. 1999a. Report on one-week visit to Cat Ba Island. Unpublished report Frankfurt Zoological Society and EPRC.

Baker, L. 1999b. Survey of the Delacour's Langur. Unpublished report Frankfurt Zoological Society and EPRC.

Ballou, J. D., Gilpin, M. and Foose, T. J. 1995. Population management for survival and recovery. New York.

BirdLife International and the Forest Inventory and Planning Institute. 2001. Sourcebook of Existing and Proposed Protected Areas in Vietnam. Hanoi.

Bleisch, W. V. and Xie Jiahua. 1998. Ecology and Behavior of the Guizhou Snub-nosed Langur (*Rhinopithecus* [*Rhinopithecus*] *brelichi*), with a Discussion of Socioecology in the Genus. In: Jablonski (ed.) The Natural History of the Doucs and Snub-nosed Monkeys. World Scientific Publishing, Singapore.

Boonratana, R. 1998. Field management of Nam Poui and Phou Xang He National Biodiversity Conservation Areas. IUCN-report. Vientiane.

Boonratana, R. 1999. Na Hang Rainforest Conservation Project. Report FFI-Indochina Programme, Hanoi.

Boonratana, R., and Le Xuan Canh. 1994. A Report on the Ecology, Status and Conservation of the Tonkin Snub-nosed Monkey (*Rhinopithecus avunculus*) in Northern Vietnam. WCS, New York and IEBR, Hanoi.

Boonratana, R., and Le Xuan Canh. 1998a. Conservation of Tonkin Snub-nosed Monkeys (*Rhinopithecus* [*Presbytiscus*] *avunculus* in Vietnam. In: Jablonski (ed.) The Natural History of the Doucs and Snub-nosed Monkeys. World Scientific Publishing, Singapore.

Boonratana, R., and Le Xuan Canh. 1998b. Preliminary Observations of the Ecology and Behaviour of the Tonkin Snub-nosed Monkey (*Rhinopithecus* [*Presbytiscus*] *avunculus*) in Northern Vietnam. In: Jablonski (ed.) The Natural History of the Doucs and Snub-nosed Monkeys. World Scientific Publishing, Singapore.

Brandon-Jones, D. 1984. Colobus and leaf monkeys. In: Macdonald, D. (ed.) The Encyclopedia of Mammals. Vol. 1. London.

Brandon-Jones, D. 1995. A revision of the Asian pied leaf monkeys (Mammalia: Cercopithecidae: superspecies *Semnopithecus auratus*), with a description of a new sub-species. *Raffles Bulletin of Zoology* 43(1): 3-43.

Brandon-Jones, D. 1996a. Further remarks on the geographic distribution and morphology *of Semnopithecus hatinhensis* and *S. francoisi* (Mammalia: Cercopithecidae) *Raffles Bulletin of Zoology* 44(1): 275-277.

Brandon-Jones, D. 1996b. The Asian Colobinae (Mammalia: Cercopithecidae) as indicators of Quaternary climatic change. *Biological Journal of the Linnean Society* 59: 327-350.

Brandon-Jones, D. 1998. Pre-glacial Bornean primate impoverishment and Wallace's line. In: Hall and Holloway (eds.) Biogeography and Geological Evolution of SE Asia. Backhuys Publishers, Leiden.

Brickle, N., Nguyen Cu, Ha Quy Quynh, Nguyen Thai Tu Cuong and Hoang Van San. 1998. The Status and Distribution of Green Peafowl, *Pavo muticus*, in Dak Lak Province, Vietnam. BirdLife International, Vietnam Programme and IEBR (Conservation Report No. 1). Hanoi.

Brotoisworo, E., and Dirgayusa, I. W. A. 1991. Ranging and feeding behaviour of *Presbytis cristata* in the Pangandaran Nature Reserve, West Java, Indonesia. In: Ehara, Kimura, Takenaka and Iwamonto (eds.) Primatology Today. Elsevier Science, Amsterdam.

Burton, F. D., Snarr, K. A., and Harrison, S. E. 1995. Preliminary Report on *Presbytis francoisi leucocephalus*. *International Journal of Primatology* 16(2): 311-329.

Cao Van Sung (ed.). 1998. Environment and Bioresources of Vietnam - Present Situation and Solutions. Hanoi.

Cao Van Sung and Pham Nhat. 1994. Ecology of Endangered Primates in Vietnam. *Chinese Primate Research and Conservation News* 3 (2): 4.

Cao Van Sung, Weitzel, V., and Vu Ngoc Thanh. 1994. Breeding and raising primates in Vietnam. *Chinese Primate Research and Conservation News* 3 (2).

Cao Van Sung, Pham Trong Anh and Le Vu Khoi. 1993. The result of biological diversity survey in Bien Lac-Nui Ong Nature Reserve. Unpublished report. (Vietnamese).

Cao Van Sung. 1994. Status of primate fauna and conservation in Vietnam. In: Xia Wuping and Zhang Yongzu (eds.) Primate Research and Conservation. China Forestry, Beijing.

Cao Van Sung. 1995. Review of the conservation status of threatened leaf monkeys in Vietnam. In: Xia Wuping and Zhang Yongzu (eds.) Primate Research and Conservation. China Forestry, Beijing.

Caton, J. M. 1998. The Morphology of the Gastrointestinal Tract of *Pygathrix nemaeus* (Linneaus, 1771). In: Jablonski (ed.) The Natural History of the Doucs and Snub-nosed Monkeys. World Scientific Publishing, Singapore.

Caughley, G. 1994. Directions in Conservation Biology. *Journ. of Animal Ecology* 63: 215-244.

Chengming Huang, Ruyong Sun and Liren Lu. 1998. The daily activity rhythm and the time budget of the White-headed Leaf Monkey (*Presbytis leucocephalus*) in Southern Guangxi, China. Proc. Int. Seminar on Commemorating the Naming of Francois' Leaf Monkey for its First Centenary and Protection the Primate 1898-1998. Wuzhou, China.

Chaplin, G., and Jablonski, N. G. 1998. The Integument of the "Odd-nosed" Colobines. In: Jablonski (ed.) The Natural History of the Doucs and Snub-nosed Monkeys. World Scientific Publishing, Singapore.

Chivers, D. J. 1973. An introduction to the socioecology of Malayan forest primates. In: Michael and Crook (eds.) Comperative Ecology and of Behavior of Primates. Academic Press, New York.

Chivers, D. J. 1994. Functional anatomy of the gastrointestinal tract. In: Davies and Oates (eds.) Colobine Monkeys: Their Ecology, Behaviour, and Evolution. Cambridge University Press, Cambridge.

Collins, N.M., Sayer, J.A., and Whitmore, T.C. (eds). 1991. The conservation atlas of tropical forests. Asia and the Pacific. Macmillan Press Ltd., London.

Compton, J. and Le Hai Quang. 1998. Borderline. An assessment of Wildlife Trade in Vietnam. Report WWF-Indochina Programme.

Conservation International and IUCN Primate Specialist Group. 2000. The world's TOP 25 most endangered Primates. Report Conservation International.

Corbet, G. B., and Hill, J. E. 1992. The Mammals of the Indomalayan Region: A Systematic Review. Natural History Museum Publications. Oxford University Press, Oxford.

Cox, R. 1994. Management Feasibility Study of the Na Hang Nature Reserve. WWF, Ministry of Forestry and IUCN, Hanoi.

Cox, C. R., Vu Van Dung and Pham Mong Giao. 1992. Report of a management feasibility study of the Muong Nhe Nature Reserve. WWF, Ministry of Forestry, Hanoi.

Cox, C. R., Vu Van Dung, Pham Mong Giao and Le Xuan Canh. 1994. A Management Feasibility Study of the Proposed Na Hang (Tonkin Snub-nosed Monkey) Nature Reserve, Tuyen Quang Province, Vietnam. NWF, IUCN, WWF Programme for Endangered Species in Asia.

Creel, N. & Preuschoft, H. 1984. Systematics of the lesser apes: A quantitative taxonomic analysis of craniometric and other variables. In: Preuschoft, Chivers, Brockelman, and Creel (eds.) The Lesser Apes: Evolutionary and Behavioural Biology. Edinburgh University Press, Edinburgh.

Dang Huy Huynh. 1995. Fauna and geographical distribution of primate species in Vietnam. In: Xia Wuping and Zhang Yongzu (eds.) Primate Research and Conservation. China Forestry, Beijing.

Dang Huy Hunh, Cao Van Sung and Le Xuan Canh. 1995. A Report on the Survey for Biological Resources in Yok Don National Park, South Vietnam. IEBR and Vietnam NCST of Vietnam, Hanoi.

Dang Huy Huynh, Dao Van Tien, Cao Van Sung, Pham Trong Anh and Hoang Minh Khien. 1994. Checklist of Mammals in Vietnam. Science and Technics Publishing House, Hanoi. (Vietnamese).

Dang Ngoc Can and Nguyen Truong Son. 1999. Field Report of a Survey on the Tonkin Snub-nosed Monkey *(Rhinopithecus avunculus)* in Bac Can, Thai Nguyen and Tuyen Quang Provinces (October and November 1999). Unpublished report FFI-Indochina Programme, Hanoi. (Vietnamese).

Dao Van Tien. 1961. Recherches zoologiques dans la région de Thai-Nguyen (Nord-Vietnam). *Zoologischer Anzeiger* 166: 298-308.

Dao Van Tien. 1970. Sur les formes de semnopithèque noir, *Presbytis francoisi* (Colobidae, Primates), au Vietnam et description d'une forme nouvelle. *Mitt. Zool. Mus. Berlin* 46: 61-65.

Dao Van Tien. 1977. Sur quelques rares mammifères au nord du Vietnam. *Mitt. Zool. Mus. Berlin* 46: 61-65.

Dao Van Tien. 1978. Sur une collection de mammifères du Plateau de Moc Chau (Province de So'n-la, Nord-Vietnam). *Mitt. Zool. Mus. Berlin* 54: 377-391.

Dao Van Tien. 1983. On the north Indochinese Gibbons (*Hylobates concolor*) (Primates: Hylobatidae) in North Vietnam. *Journal of Human Evolution* 12: 367-372.

Dao Van Tien. 1985. Scientific Results of Some Mammals Surveys in North Vietnam (1957-1971). Scientific and Technics Publishing House, Hanoi. (Vietnamese, English summary).

Dao Van Tien. 1989. On the trends of the evolutionary radiation on the Tonkin Leafmonkey (*Presbytis francoisi*, primates: Cercopithecidae). *Journal of Human Evolution* 4: 501-507.

Dao Van Tien. Undated. On two Leaf-monkeys (Presbytis sp.) considered threatened (T) in Vietnam. MS, Hanoi.

Dao Van Tien and Tran Hong Viet. 1984. Mammal checklist in Sa Thay District. *Journal of Biology* 6(2): 28-30. (Vietnamese).

Davies, G. A., and Oates, J. F. (eds.) 1994. Colobine Monkeys: Their Ecology, Behaviour and Evolution. Cambridge University Press, Cambridge.

Davis, S. D., Heywood, V. H., and Hamilton, C. H. 1995. Centres of Plant Diversity, A Guide and Strategy for their Conservation. Vol 2 Asia, Australasia and the Pacific. WWF and IUCN.

Dawson, S., Do Tuoc, Le Vu Khoi and Trinh Viet Cuong. 1993. Elephant surveys in Vietnam. Report WWF-Vietnam Programme, Hanoi.

Delson, E. 1994. Evolutionary history of the colobine monkeys in paloenvironmental perspective. In: Davies and Oates (eds.) Colobine Monkeys: Their Ecology, Behaviour, and Evolution. Cambridge University Press, Cambridge.

Dene, H. T., Goodman, M., and Prychodko, W. 1976. Immunodiffusion evidence on the phylogeny of the primates. In: Goodman, Tashian and Tashian (eds.) Molecular anthropology. Genes and proteins in the evolutionary ascent of the primates. Plenum Press, New York.

Desai, A. A., and Vuthy, L. 1996. Status and distribution of large mammals in eastern Cambodia. Results of the first foot surveys in Mondulkiri and Rattanakiri Provinces. IUCN/FFI/WWF Large Mammal Conservation Project, Phnom Penh, Cambodia.

Dinh Trong Thu. 1998. Socio-economic research on dependency by the people on forest resources. Na Hang Rain Forest Conservation Project, Report FFI-Indochina Programme, Hanoi.

Do Tuoc . 1995. Features and Values of Fauna Component of the Chu Mom Ray Nature Reserve. FIPI, Hanoi. (Vietnamese).

Dolhinow, P. and DeMay, M. G. 1982. Adoption - the importance of infant choice. *Journ. of Human Evolution* 11: 391-420.

Dollman, G. 1912. A new snub-nosed monkey. *Proceedings of the Zoological Society of London 1912.* 106: 503-504.

Dong Thanh Hai and Lormée, N. 1999. Status Assessment of the Black Gibbon (*Hylobates concolor concolor*) in Van Ban District, Lao Cai Province, North Vietnam, 14th-28th November *1999*. Unpublished report FFI-Indochina Programme, Hanoi.

Duckworth, J. W., Salter, R. E., and Khounboline, K. (eds.). 1999. Wildlife in Lao PDR, 1999 Status Report. IUCN, WCS and CPAWM, Vientiane.

Duckworth, J. W., Timmins, R. J., Khounboline, K., Salter, R. E., and Davidson, P. 1999. Large mammals. In: Duckworth, Salter, and Khounboline (eds.) Wildlife in Lao PDR: 1999 status report. IUCN – The World Conservation Union / Wildlife Conservation Society / Centre for Protected Areas and Watershed Management, Vientiane.

Duckworth, J. W., and Walston, J. L. 1998. Reconnaissance bird and mammal survey of Hai Phong - Ha Long - Cat Ba - Cam Pha - Ba Mun area, Hai Phong and Quang Ninh Provinces, Vietnam, with emphasis on the endemic langur. Unpublished report World Bank.

Eames, J. C., and Nguyen Cu. 1994. A management feasibility study of Thuong Da Nhim and Chu Yang Sin Nature Reserves on the Da Lat Plateau, Vietnam. WWF-Vietnam Programme, Hanoi.

Eames, J. C., and Robson, C. R. 1993. Threatened primates in southern Vietnam. *Oryx* 27: 146-154.

Eames, J. C., Robson, C. R., Nguyen Cu and Truong Van La. 1991. Forest Birds Surveys in Vietnam. International Council for Bird Preservation (Study Report no. 51).

Eames, J. C., Kuznetsov, A. N., Monastyrskii, A. L., Nguyen Tien Hiep, Nguyen Quang Truong and Ha Quy Quynh. 2001. A Preliminary Biological Assessment of Kon Plong Forest Complex, Kon Tum Province, Vietnam. WWF-Indochina Programme, Hanoi.

Ellerman, J. T. and Morrison-Scott, T. C. S. 1951. Checklist of Palearctic and Indian mammals, 1758-1946. Trustees of the British Museum, London.

Elliot, D. G. 1909. Descriptions of apparently new species and sub-species of monkeys of the genera *Callicebus, Lagothrix, Papio, Pithecus, Cercopithecus, Erythrocebus*, and *Presbytis. Ann. Mag. Nat. Hist.* (8th ser.) 4: 244-274.

Eudey, A. A. 1987. Action Plan for Asian Primate Conservation *1987-1991*. IUCN-SSC Primate Specialist Group, IUCN, Gland, Switzerland and Cambridge.

Eudey, A. A. 1996/1997. Asian primate conservation – The species and the IUCN/SSC primate specialist network. *Primate Conservation* 17: 101-110.

Eve, R., Nguyen Viet Dung and Meijboom, M. 1998. Vu Quang Nature Reserve. A Link in the Annamite Chain. Vol. 2, No 0 : List of species-Fauna and Flora. DGIS (Activity no. VN 003301) and WWF (Project no. VN 0021).

Felsenstein, J. 1987. Estimation of hominoid phylogeny from a DNA hybridization data set. *Journal of Molecular Evolution* 26: 123-131.

Fleagle, J. G. 1978. Locomotion, posture, and habitat utilization in two sympatric Malaysian leaf-monkeys (*Presbytis obscura* and *Presbytis melalophos*). In: Montgomery G. G. (ed.) The Ecology of Arboreal Folivores. Smithsonian Institution Press, Washington DC.

Fleagle, J. G. 1999. Primate adaptation and evolution. Academic Press, San Diego and London.

Fooden, J. 1971. Report of primates collected in western Thailand. *Fieldiana Zool.* 59: 1-62.

Fooden, J. 1975. Taxonomy and evolution of liontail and pigtail macaques (Primates: Cercopithecidae). *Fieldiana Zool.* 67: 1-169.

Fooden, J. 1976a. Provisional classification and key to living species of macaques Primates: *Macaca*. *Folia Primatologica* 25: 225-236.

Fooden, J. 1976b. Primates obtained in peninsular Thailand June-July 1973, with notes on the distribution of continental southeast Asian leaf-monkeys (*Presbytis*). *Primates* 17: 95-118.

Fooden, J. 1996. Zoogeography of Vietnamese Primates. *International Journal of Primatology* 17(5): 845-899.

Geissmann, T. and Vu Ngoc Thanh. 2000. Preliminary results of a primate survey in northeastern Vietnam, with special reference to gibbons. *Asian Primates* 7 (3): 1-4.

Geissmann, T., Nguyen Xuan Dang, Lormée, N. and Momberg, F. 2000. Vietnam Primate Conservation Status Review 2000. Part 1: Gibbons. Fauna & Flora International-Indochina Programme, Hanoi.

Geoffroy Saint-Hillaire, E. 1812. Tableau des quadrumanes, ou des animaux composant le premier Ordre de la Classe des Mammifères. *Annls Mus. Hist. Nat. Paris* 19: 85-122.

Gerhardt, U. 1909. Ueber das Vorkommen eines Penis- und Clitorisknochens bei Hylobatiden. *Anatomischer Anzeiger* 35: 353-358.

Ghazoul, J., and Le Mong Chan. 1994. Nui Hoang Lien Nature Reserve. SEE Vietnam Forest Research Programme (Technical Report no. 2), Hanoi and SEE, London.

Ghazoul, J., Le Mong Chan and Listen, K. 1994. Scientific Report for Ba Na Nature Reserve, Vietnam. SEE, London, Ministry of Forestry Hanoi and Department of Forestry Danang.

Gibson, D., and Chu, E. 1992. Management and behaviour of Francois' langur *Presbytis francoisi francoisi* at the Zoological Society of San Diego. *Int. Zoo Yb.* 31: 184-191.

Gilmour, D. A., and Nguyen Van San. 1999. Buffer zone management in Vietnam. IUCN Vietnam, Hanoi.

Gochfeld, M. 1974. Douc langurs. *Nature* 247: 167.

Goldman, D., Giri, P. R., and O'Brien, S. J. 1987. A molecular phylogeny of the hominoid primates as indicated by two-dimensional protein electrophoresis. *Proceedings of the National Academy of Sciences USA* 84: 3307-3311.

Goodman, M., Porter, C. A., Czelusniak, J., Page, S. L., Schneider, H., Shoshani, J., Gunnell, G., and Groves, C. P. 1998. Toward a phylogenetic classification of primates based on DNA evidence complemented by fossil evidence. *Molecular Phylogenetics and Evolution* 9: 585-598.

Goodman, M., Tagle, D. A., Fitch, D. H. A., Bailey, W., Czelusniak, J., Koop, B. F., Benson, P., and Slightom, J. L. 1990. Primate evolution at the DNA level and a classification of hominoids. *Journal of Molecular Evolution* 30: 260-266.

Government of the Socialist Republic of Vietnam and Global Environment Facility Project VIE/91/G31. 1994. Biodiversity Action Plan for Vietnam. Hanoi.

Groves, C. P. 1970. The forgotten leaf-eaters, and the phylogeny of the Colobinae. In: Napier, J. R., and Napier, P. H. (eds.) Old World Monkeys: Evolution, Systematics, and Behaviour. Academic Press, New York.

Groves, C. P. 1989. A Theory of Human and Primate Evolution. Clarendon Press, Oxford.

Groves, C. P. 1992. Order Primates. In: Wilson and Reeder (eds.) Mammal Species of the World. Smithsonian Institution Press, Washington DC.

Groves, C. P. 1993. Speciation in living hominoid primates. In: Kimbel and Martin (eds.) Species, species concepts, and primate evolution. New York and London, Plenum Press.

Groves, C. P. 2001. Primate taxonomy. Smithsonian Institution Press, Washington D.C.

Ha Dinh Duc. 1995. Hatinh Monkey (Trachypithecus francoisi hatinhensis) endemic sub-species of North Truong Son. Selected papers of seminar on North Truon Son Biodiversity. 1995:100-104. Science and Technics Publishing House, Hanoi.

Ha Thang Long. 1999. Distribution of Delacour's langur in Ba Thuoc, Kim Bang and Gia Vien districts. Unpublished report Frankfurt Zoological Society.

Ha Thang Long. 2000. Douc langur Survey in Central and South Vietnam - May to July and October 2000. Unpublished report Frankfurt Zoological Society.

Ha Thang Long. 2001. Primate Survey Report with special emphasis on the Black-shanked Douc langur (Pygathrix nigripes) in Ninh Thuan and Binh Thuan provinces, South Vietnam. Unpublished report Frankfurt Zoological Society.

Ha Thang Long. 2002. Primate Survey Report with special emphasis on the Black-shanked Douc langur (Pygathrix nigripes) in Lam Dong Province, South Vietnam. Unpublished report Frankfurt Zoological Society.

Ha Thang Long and Le Thien Duc. 2001. Primate Survey Report with special emphasis of the Black-shanked Douc langurs (Pygathrix nigripes) in Binh Phuoc and Dak Lak Provinces, South Vietnam. Unpublished report Frankfurt Zoological Society.

Harlan, R. 1826. Description of an hermaphrodite orang outang lately living in Philadelphia. J. Acad. Nat. Sci. Philadelphia 5: 229-236.

Harding, J. and Groves, C. P. 2001. Craniometric studies of the Douc langurs (Pygathrix). Proc. XVIII. Congress of Int. Primat. Soc.: 382. Adelaide 2001.

Hayssen, V., van Tienhoven, A. and van Tienhoven, A. 1993. Asdell's patterns of mammalian reproduction: a compendium of species-specific data. Comstock/Cornell University Press. Ithaca.

Hill, M. 1997. Primates in protected areas of Northern Vietnam. SEE- report, London.

Hill, M. 1999. Mammals of Cuc Phuong National Park. Unpublished report.

Hill, M., and Hallam, D. 1997. Na Hang Nature Reserve, Part 2: Tat Ke sector. Biodiversity survey 1996. SEE-Vietnam report no. 7. Hanoi and London.

Hill, M., Hallam, D., and Bradeley, J. 1996a. Ba Be National Park -Biodiversity survey 1996. SEE Vietnam Forest Research Programme Technical report no. 10. Hanoi and London.

Hill, M., Hallam, D., and Bradeley, J. 1997. Muong Nhe Nature Reserve Biodiversity survey 1997. Frontier Vietnam Forest Research Programme Technical Report No. 11. Hanoi and London.

Hill, M., Kemp, N., Dung Ngoc Can, Truong Van La and Ha Van Tue. 1996a. Biological survey of Na Hang Nature Reserve, Tuyen Quang Province, Vietnam. Part 1 Ban Bung sector. SEE-Vietnam Technical Report No. 1. Hanoi and London.

Hill, M., Nguyen Kiem Son, Le Mong Chan and Harrisson, E. M. 1996b. Site Description and Conservation Evaluation: Ba Na Forest Reserve, Quang Nam-Da Nang Province, Vietnam. Frontier, Vietnam Forest Research Programme. Hanoi and London.

Hill, W. C. O. 1934. A monograph on the purple-faced leaf-monkeys (*Pithecus vetulus*). *Ceylon Journal of Science* (B) 9: 23-88.

Hilton-Taylor, C. (compiler) 2000. 2000 IUCN Red List of Threatened Species. IUCN, Gland and Cambridge.

Hoang Hoe, Le Van Lanh, Nguyen Ba Thu, Nguyen Duc Khang and Vu Van Dung. 2001. National Parks of Vietnam. Hanoi. (Vietnamese).

Hrdy, S. B. 1976. The care and exploitation of non-human primate infants by con-specifics other than the mother. In: Rosenblatt Hinde, Shaw, and Beer (eds.) Advances in the Study and Behaviour; Vol. 6. Academy Press, New York.

Hrdy, S. B. 1977. The langurs of Abu: Female and male strategies of reproduction. Harvard University Press. Cambridge.

Huang Chengming, Sun Ruyong and Lu Liren. 1998. The daily activity rhythm and the time budget of White-headed Leaf Monkey (*Presbytis leucocephalus*) in Southern Guangxi, China. Proc. Int. Seminar on Commemorating the Naming of Francois' Leaf Monkey for its First Centenary and Protection the Primate. Wuzhou, China.

Huynh Van Keo and Van Ngoc Thinh. 1998. The Status of Douc Langur (*Pygathrix nemaeus*) and some Preliminary Results of Wildlife Conservation in Bach Ma National Park. Proceedings Workshop on a Conservation Action Plan for the Primates of Vietnam. Hanoi.

Jablonski, N. G. 1995. The Phyletic Position and Systematics of the Douc Langurs of Southeast Asia. *American Journal of Primatology* 35: 185-205.

Jablonski, N. G. 1998. The evolution of the Doucs and Snub-nosed Monkeys and the Question of the Phyletic Unity of the Odd-nosed Colobines. In: Jablonski (ed.) The Natural History of the Doucs and Snub-nosed Monkeys. World Scientific Publishing, Singapore.

Jablonski, N. G. and Pan Ruliang. 1995. Sexual Dimorphism in the Snub-Nosed Langurs (Colobinae: *Rhinopithecus*). *American Journal of Physical Anthropology* 96: 251-272.

Jablonski, N. G. and Peng Yanzhang. 1993. The Phylogenetic Relationships and Classification of the Doucs and Snub-Nosed Langurs of China and Vietnam. *Folia Primatologica* 60: 36-55.

Jablonski, N. G., Pan Ruliang and Chaplin, G. 1998. Mandibular Morphology of the Doucs and Snub-nosed Monkeys in Relation to Diet. In: Jablonski (ed.) The Natural History of the Doucs and Snub-nosed Monkeys. World Scientific Publishing, Singapore.

Johns, B. G. 1999. A survey programme to establish conservation and management priorities for the primates of the Pu Mat Nature Reserve, Nghe An Province, Vietnam. SFNC in Nghe An Province. Vinh.

Kavanagh, M. 1972. Food-sharing behaviour within a group of Douc monkeys *Pygathrix nemaeus nemaeus*. *Nature* 239: 406-407.

Kemp, N., and Dilger, M. 1996. Site Description and Conservation Evaluation: Bu Huong Proposed Nature Reserve Quy Chau District, Nghe An Province, Vietnam. Frontier, Vietnam Forest Research Programme, Scientific Report No. 7. Hanoi and London.

Kemp, N., Le Mong Chan & Dilger, M. 1994. Site Description and Conservation Evaluation: Ba Be National Park, Cao Bang Province, Vietnam. Frontier, Vietnam Forest Research Programme, Scientific Report No. 4. Hanoi and London.

Kemp, N., Le Mong Chan and Dilger, M. 1995b. Site Description and Conservation Evaluation: Pu Mat Nature Reserve, Con Cuong District, Nghe An Province, Vietnam. Frontier, Vietnam Forest Research Programme, Technical Report No. 5. Hanoi and London.

Kirkpatrick, R. C. 1998. Ecology and Behaviour in Snub-nosed and Douc langurs. In: Jablonski (ed.) The Natural History of the Doucs and Snub-nosed Monkeys. World Scientific Publishing, Singapore.

Kirkpatrick, R. C. 1998. Toward a Gazetteer of the Snub-nosed and Douc Langurs. In: Jablonski (ed.) The Natural History of the Doucs and Snub-nosed Monkeys. World Scientific Publishing, Singapore.

Kloss, C. B. 1916. On a collection of Mammals form the Coast and Islands of South-East Asia. *Proceedings of the Zoological Society of London* 1916: 25-75.

Kloss, C. B. 1919. On Birds from South Annam and Cochin China. Part I. Phasianidae-Campophagidae. Narrative of the Journey. *Ibis* 1: 392-402.

Kloss, C. B. 1919. On Mammals Collected in Siam. *J. Nat. Hist. Soc. Siam* 3: 333-407.

Kloss, C. B. 1926. A new Race of Monkey from Annam. *Ann. Mag. Nat. Hist.* (9th ser.) 18: 214.

Kloss, C. B. 1929. Some remarks on the gibbons with a new sub-species. *Proceedings of the Zoological Society of London*: 113-127.

Koontz, F. W. 1997. Zoos and in situ Primate Conservation. In: J. Wallis (ed.) Primate Conservation: The Role of Zoological Parks. Oklahoma.

La Dang Bat. 1998. Vooc quan dui trang co o Ninh Binh (Delacour's langur in Ninh Binh). Lao Dong 22/98 (7. 2. 1998). (Vietnamese).

La Quang Trung & Trinh Dinh Hoang. 2001. The Eastern black gibbon *(Nomascus* sp. cf. *nasutus)* survey in Kim Hy Nature Reserve, Na Ri District, Bac Kan Province, June 2001. Report Fauna & Flora International, Indochina-Programme. Hanoi.

La Quang Trung. 2001. Primate focused surveys in Ha Lang District, Cao Bang province. Report Fauna & Flora International, Indochina-Programme. Hanoi.

La Quang Trung and Trinh Dinh Hoang. 2002. Rapid assessment for Northern White-cheeked Gibbons and other primate species in Xuan Lien Nature Reserve and adjacent forests in Pu Huot proposed Nature Reserve. Report Fauna & Flora International, Indochina-Programme. Hanoi.

Lambert, F. 1995. Endangered species in South-east Asia. *IUCN Bull.* 26(3): 25-26.

Lambert, F. A., Eames, J. C., and Nguyen Cu. 1994. Surveys for Endemic Pheasants in the Annamese Lowlands of Vietnam, June-July, 1994. NWF, IUCN, WWF Programme for Endangered Species in Asia.

Lapin, B. A., Dzhikidze, E. K., and Jakovleva, L. A. 1965. Diseases of Laboratory Primates under Natural Conditions in Vietnam. Meditsina, Moscow. (Russian).

Le Hien Hao. 1973. Economic value of Mammals in North Vietnam. Scientific and Technical Publishing House, Hanoi. (Vietnamese).

Le Khac Quyet. 2001. Exploratory Survey of Wildlife in Ha Giang Province. Report Fauna & Flora International, Indochina-Programme, Hanoi.

Le Khac Quyet and La Quang Trang. 2001. A Preliminary Survey on Primates in Cao Bang Province. Report Fauna & Flora International, Indochina-Programme, Hanoi

Le Khac Quyet and Trinh Dinh Hoang. 2003a. A Survey on Primate Fauna and Assessment of Conservation Potential in the Ban Thi-Xuan Lac and Tan Lap Areas, Cho Don District, Bac Kan Province. Conservation Report Fauna & Flora International, Vietnam Programme/ PARC Project.

Le Khac Quyet and Trinh Dinh Hoang. 2003b. Field Surveys and Assessments In Duc Xuan-Sinh Long and Lung Nhoi Areas, Na Hang District, Tuyen Quang Province, with special Reference to Primates. Conservation Report Fauna & Flora International, Vietnam Programme/ PARC Project.

Le, T. 1997. Report on Animal Resources of Pu Luong Nature Reserve, Thanh Hoa Province. FIPI report, Hanoi. (Vietnamese).

Le Thien Duc. 2002. Survey report for Delacour's langurs in Nui Ke mountainous area. Unpublished report Frankfurt Zoological Society.

Le Trong Dat, La Quang Trung and Trinh Dinh Hoang. 2000. A biological survey of Nam Xay and Nam Xe communes, Van Ban District, Lao Cai Province with specific focus on the western black crested gibbon. Report Fauna & Flora International, Indochina-Programme, Hanoi.

Le Trong Trai. 1999a. A Feasibility Study for the Establishment of Xuan Lien Nature Reserve, Thanh Hoa Province, Vietnam. BirdLife International, Vietnam Programme and FIPI (Conservation Report 7), Hanoi.

Le Trong Trai. 1999b. Fauna and Flora of Kon Chu Rang Nature Reserve, Gia Lai Province. MARD and FIPI, Hanoi. (Vietnamese).

Le Trong Trai. 2000. An Investment Plan for Kon Ka Kinh Nature Reserve, Gia Lai Province, Vietnam. A Contribution to the Management Plan. BirdLife International, Vietnam Programme and FIPI (Conservation Report No. 11), Hanoi.

Le Trong Trai, Nguyen Cu, Le Van Cham, Eames, J. C., and Tran Van Khoa. 1996. Biodiversity study to consider the feasibility to extend Chu Yang Sin Nature Reserve, Dac Lac Province. MARD and People's Committe of Dak Lak Province, Buon Ma Thuot. (Vietnamese).

Le Trong Trai, Nguyen Huy Dung, Nguyen Cu, Le Van Cham, Eames, J. C., and Chicoine, G. 1996. An Investment Plan for Ke Go Nature Reserve, Ha Tinh Province, Vietnam. A Contribution to the Management Plan. BirdLife International, Vietnam Programme and FIPI (Conservation Report No. 9), Hanoi.

Le Trong Trai, Eames, J. C., Kutznetsov, A. N., Nguyen Van Sang, Hayes, B. D., Nguyen Truong Son, Bui Xuan Phuong, Monastirskii, A. L. and Tordorff, A. W. 2001. A Biodiversity Survey of the Dong Phuc, Ban Thi-Xuan Lac, and Sinh Long areas: PARC Project Na Hang/Ba Be Component, Hanoi.

Le Trong Trai and Richardson, W. J. 1999a. A Feasibility Study for the Establishment of Phong Dien (Thua Thien Hue Province) and Dakrong (Quang Tri Province) Nature Reserves, Vietnam. BirdLife International, Vietnam Programme and FIPI (Conservation Report No. 4), Hanoi.

Le Trong Trai and Richardson, W. J. 1999b. An Investment Plan for Ngoc Linh Nature Reserve, Kon Tum Province, Vietnam. A Contribution to the Management Plan. BirdLife International, Vietnam Programme and FIPI (Conservation Report No. 5), Hanoi.

Le Xuan Canh. 1993. Evidence for the existence of *Trachypithecus francoisi hatinhensis*. *Asian Primates* 2(3-4): 2.

Le Xuan Canh. 1994. New Information about the Tonkin Snub-nosed Monkey, *Rhinopithecus avunculus*, in Na Hang forest. Unpublished Manuscript.

Le Xuan Canh. 1995a. A report on the survey for large carnivores in Tay Nguyen Plateau, South Vietnam with emphasis on tiger *(Panthera tigris)*. WCS, Hanoi.

Le Xuan Canh. 1995b. Biology, ecology and population structure and number of Tonkin snub-nosed monkey *(Rhinopithecus avunculus)* in Vietnam. In: Xia Wuping and Zhang Yongsu (eds.) Primate Research and Conservation. China Forestry, Beijing.

Le Xuan Canh, Myers, M., Rowe, N., Weitzel, V., and Le Hong Binh. 1993. Survey for the Tonkin Snub-nosed Monkey *(Rhinopithecus avunculus)*. Unpublished report to Primate Conservation, Inc.

Le Xuan Canh and Campbell, B. 1994. Population status of Golden-headed Langur *(Trachypithecus francoisi poliocephalus)* in Cat Ba National Park. *Asian Primates* 3(3-4): 16-20.

Le Xuan Canh, Pham Trong Anh, Duckworth, J. W., Vu Ngoc Thanh and Vuthy, L. 1997a. A Survey of Large Mammals in Dak Lak province, Vietnam. WWF and IUCN, Hanoi.

Le Xuan Canh, Truong Van La, Dang Thi Dap, Ho Thu Cuc, Ngo Anh Dao, Nguyen Ngoc Chinh, Nguyen Quoc Dung, Pham Nhat, Nguyen Thai Tu, Nguyen Quoc Thang and Tran Minh Hien. 1997b. A Report on Field surveys on Biodiversity in Phong Nha–Ke Bang Forest (Quang Binh Province) Central Vietnam. WWF and UNDP, Hanoi.

Le Xuan Canh, Cao Van Sung and Sang Don Lee. 1997c. Mammal Resources of Cat Ba and Surrounding Areas in Vietnam. In: Ecosystem and Biodiversity of Cat Ba National Park and Halong Bay, Vietnam. Ann. of Nature Conservation, Korean National Council for Conservation of nature, vol. 12: Survey of the Natural environment in Vietnam.

Le Xuan Canh, Hoang Minh Khien, Le Dinh Thuy, Ho Thu Cuc, Hoang Vu Tru, Ngo Van Tri, Nguyen Tran Vy and Tran Viet Dung. 1998. Results of a wildlife survey in Cat Loc Nature Reserve. Unpublished report IEBR, Hanoi and ITB, Ho Chi Minh City. (Vietnamese).

Le Xuan Canh, Do Huu Thu and Nguyen Van San. 2000. Results of survey on biodiversity and socio-economic conditions in Cham Chu mountain area, Tuyen Quang Province. Unpublished report Institute of Ecology and Biological Resources, Hanoi. (Vietnamese).

Lekagul, B., and McNeely, J. A. 1977. Mammals of Thailand. Association for the Conservation of Wildlife, Bangkok.

Lewis, N. 1951. A Dragon Apparent. Travels in Cambodia, Lao PDR, and Vietnam. Eland Books, London.

Li Wenjun, Fuller, T. K., and Wang Sung. 1996. A survey of wildlife trade in Guangxi and Guangdong, China. *Traffic Bulletin* 16 (1): 9-16.

Li, Z. X. and Ma, S. L. 1980. Taxonomic review on the White-headed langur. *Acta Zootax. Sin.* 5, 440-442.

Li Zhaoyuan, Liu Zimin, Wei Yi and Rogers, E. in press. Decline in Population Density of the White-headed Langur, *Presbytis leucocephalus*, in Guangxi, China.

Li Zhaoyuan. 1999. Study of the Conservation Biology on the White-headed Langur, *Prebytis leucocephalus*, in China. *ASP Bull.* 23 (2), 6.

Ling, S. 2000. A survey of wild cattle and other mammals, Cat Tien National Park. WWF-Cat Tien Conservation Project (Technical Report No. 14).

Lippold, L. K. 1977. The Douc Langur: A Time for Conservation. In: Prince Rainier III and Bourne, G. H. (eds.) Primate Conservation. Academic Press, New York.

Lippold, L. K. 1995a. Distribution and conservation of the Douc Langur (*Pygathrix nemaeus*) in Vietnam. In: Xia Wuping and Zhang Yongsu (eds.) Primate Research and Conservation. China Forestry, Beijing.

Lippold, L. K. 1995b. Distribution and conservation status of Douc langurs in Vietnam. *Asian Primates* 4(4): 4-6.

Lippold, L. K. 1998. Natural History of Douc Langurs. In: Jablonski (ed.) The Natural History of the Doucs and Snub-nosed Monkeys. World Scientific Publishing, Singapore.

Lippold, L. K., and Vu Ngoc Thanh. 1995a. A new location for *Trachypithecus francoisi hatinhensis*. *Asian Primates* 4(4): 6-8.

Lippold, L. K., and Vu Ngoc Thanh. 1995b. Douc langur variety in the central highlands of Vietnam. *Asian Primates* 5(1-2): 6-8.

Lippold, L. K., and Vu Ngoc Thanh. 1998. Primate Conservation in Vietnam. In: Jablonsky (ed.) The Natural History of the Doucs and Snub-nosed Monkeys. World Scientific Publishing, Singapore.

Lippold, L. K., and Vu Ngoc Thanh. 1999. Distribution of the grey shanked douc langur in Vietnam. *Asian Primates* 7(1-2): 1-3.

Long, B., Le Khac Quyet and Le Trong Dat. 2000a. The mammalian and avian diversity of Che Tao commune, Mu Cang Chai district, Yen Bai province. Report Fauna & Flora International, Indochina-Programme, Hanoi.

Long, B., Bunthoeun R., Holden, J. and Seiha, U. 2000b. Large Mammals In: Daltry, J. C. and Momberg, F. (eds.) Cardamom Mountains; Biodiversity survey 2000. Fauna & Flora International, Cambridge, UK.

Long, B., Swan, S. R. and Masphal, K. 2000c. Biological Surveys in North East Mondulkiri, Cambodia. Report Fauna & Flora International, Indochina-Programme, Hanoi and Phnom Penh.

Long, B. and Swan, S. R. 2000. Cambodian Primates: Surveys in the Cardamom Mountains and North east Mondulkiri Province. Report Fauna & Flora International, Indochina-Programme, Hanoi and Phnom Penh.

Long, B. and Le Khac Quyet. 2001. An initial assessment of conservation requirements for Cham Chu, Tuyen Quang province including mammal and bird diversity surveys. Unpubl. report Fauna & Flora International, Indochina-Programme, Hanoi.

Long, B., Tallents, L. A. and Tran Dinh Nghia. 2001. The biological diversity of the Che Tao forests, Mu Cang Chai District, Yen Bai Province and Muong La District, Son La Province, Vietnam. Fauna & Flora International, Indochina-Programme, Hanoi.

Lowe, W. P. 1947. The End of the Trail. James Townsend and Sons, Exeter.

Luong Van Hao. 1999a. Report about investigation and survey of the Delacour's langur. Unpublished Report Frankfurt Zoological Society. (Vietnamese).

Luong Van Hao. 1999b. Report about distribution of the Delacour's langur in Hoa Binh province. Unpublished Report Frankfurt Zoological Society. (Vietnamese).

Luong Van Hao. 2000a. Report about distribution and ecology of the Delacour's langur. Unpublished Report Frankfurt Zoological Society. (Vietnamese).

Luong Van Hao. 2000b. Report about investigation and survey of the Delacour's langur in Cuc Phuong National Park. Unpublished Report Frankfurt Zoological Society. (Vietnamese).

Luong Van Hao. 2000c. Report about distribution and ecology of the Delacour's langur. Unpublished Report Frankfurt Zoological Society. (Vietnamese).

MacKinnon, J., and MacKinnon, K. 1987. Conservation status of the primates of the Indo-Chinese subregion. *Primate Conservation* 8: 187-195.

MacKinnon, J., Laurie, A., Yok Don Nhieu, Dang Huy Huynh, Le Khoi and Ha Dinh Duc. 1989. Draft Management Plan for Yok Don Nature Reserve Easup District, Daklak Province, Vietnam. WWF-Hong Kong.

MacKinnon, J. 1992. Draft management plan for Vu Quang Nature Reserve, Huong Khe District, Ha Tinh Province, Vietnam. Unpublished manuscript. WWF and IEBR, Hanoi.

MacKinnon, K. 1986. The conservation status of nonhuman primates in Indonesia. In: Benirschke, K. Primates: The Road of Self-Sustaining population. Springer-Verlag, New York.

Marsh, C. W. and Wilson, W. W. 1981. A Survey of Primates in Peninsular Malaysian Forests. Universiti Kebangsaan, Kuala Lumpur.

Martin, B. 2000. Tonkin snub-nosed monkey Conservation Project (TCP). Report on the first phase of the project from December 1997 until March 2000. Hanoi. (in print).

Martin, E. B., and Phipps, M. 1996. A Review of the Wild Animal Trade in Cambodia. *TRAFFIC Bulletin,* Vol.16 (2): 45-60.

Mei Qu Nian, Lai Maoqing and Li Guoqin 1998. The feeding of the *Presbytis francoisi* and Conservation of its population. Proc. Int. Seminar on Commemorating the Naming of Francois' Leaf Monkey for its First Centenary and Protection the Primate. Wuzhou, China.

Mei Qu Nian. 1998. The growth of *Presbytis francoisi.* Proc. Int. Seminar on Commemorating the Naming of Francois' Leaf Monkey for its First Centenary and Protection the Primate. Wuzhou, China.

Mey, E. 1994. Pedicinus-Formen seltener Schlankaffen aus Vietnam. *Rudolstaedter nat. hist. Schriften* 6, 83-92.

Ministry of Science, Technology and Environment. 1992. Red Data Book of Vietnam. Vol. 1: Animals. Science and Technics Publishing House, Hanoi. (Vietnamese).

Milne-Edwards, A. 1871. Note sur une nouvelle espèce de semnopithèque provenant de la Cochinchine. *Bull. Nouv. Arch. Mus.* 6: 7-9.

Milne-Edwards, A. 1877. Sur quelques mammifères et crustacés nouveaux. *Bull. Soc. Philom. Paris* 13 (1876): 8 –10.

Milne-Edwards, A. 1897. Note sur une nouvelle espèce du genre Rhinopithèque. *Bulletin Mus. Hist., Nat. Paris* 3: 156-160.

Mitani, J. C. 1990. Experimental field studies of Asian ape social systems. *International Journal of Primatology* 11: 103-126.

Momberg, F. & Fredrickson, G. 2003. A Species Recuperation and Action Plan for Tonkin Snub-nosed Monkey and Francois' Langur in Tuyen Quang and Bac Kan Provinces, Vietnam. Conservation Report, Fauna & Flora International, Vietnam Programme, PARC Project.

Monestrol, H. (ed.). 1952. Chasses et faune d'Indochine. Saigon.

Morice, A. 1875. Coup d'oeil sur la faune de la Cochinchine francaise. H. Georg, Lyon, France.

Morice, A. 1876. Voyage en Cochinchine pendant les années 1872-73-74. H. George, Lyon, France.

Mouhot, H. 1864. Travel in the Central Parts of Indo-China (Siam), Cambodia and Lao PDR, during the Years 1858, 1859 and 1860, Vol. I and II. John Murray, London.

Nadler, T. 1994. Zur Haltung von Delacour- und Tonkin-Languren (*Trachypithecus delacouri* und *Trachypithecus francoisi*) im Gebiet ihres natuerlichen Lebensraumes. *Zoologische Garten* (N.F.) 64(6): 379-398.

Nadler, T. 1995a. Checklist of Mammals of the Cuc Phuong National Park. Report Frankfurt Zoological Society, Cuc Phuong National Park Conservation Programme.

Nadler, T. 1995b. Douc langur *Pygathrix nemaeus* ssp. and Francois' langur *Trachypithecus francoisi* ssp. with questionable taxonomic status in the Endangered Primate Rescue Center, Vietnam. *Asian Primates* 5(1-2): 8-10.

Nadler, T. 1995c. Distribution and status of Delacour's and Tonkin langur in Vietnam and their taxonomic position. Proceedings 69. Tagung der Deutschen Gesellschaft fuer Saeugetierkunde, Goettingen.

Nadler, T. 1996a. Kleideraffe und Tonkinlangur (*Pygathrix* und *Trachypithecus*) mit fraglichem taxonomischen Status im "Endangered Primate Rescue Center". *Mitteilungen Zoologische Gesellschaft fuer Arten- und Populationsschutz* 12(1): 1-3.

Nadler, T. 1996b. Report on the distribution and status of Delacour's langur (*Trachypithecus delacouri*). *Asian Primates* 6(1-2): 1-4.

Nadler, T. 1996c. Endangered Primate Rescue Center: Report 1993-1995. *Endangered Primate Rescue Center Newsletter* No. 1, 4-15.

Nadler, T. 1996d. Delacour- und Hatinh langur (*Trachypithecus delacouri* und *T. francoisi hatinhensis*) erstmals in einer Haltung geboren. *Mitteilungen Zoologische Gesellschaft fuer Arten- und Populationsschutz* 12(2): 6-7.

Nadler, T. 1996e Verbreitung und Status von Delacour-, Tonkin- und Goldschopf-languren (*Trachypithecus delacouri*, *Trachypithecus francoisi* und *Trachypithecus poliocephalus*) in Vietnam. *Zoologische Garten* (N.F.) 66(1): 1-12.

Nadler, T. 1997a. A new sub-species of Douc langur, *Pygathrix nemaeus cinereus* ssp. nov. *Zoologische Garten* (N.F.) 67(4): 165-176.

Nadler, T. 1997b. Aufzucht und Jugendentwicklung von Delacour- (*Trachypitheaus delacouri*) und Hatinh-Languren (*Trachypithecus francoisi hatinhensis*). *Zoologische Garten* (N.F.) 67(4): 201-219.

Nadler, T. 1997c. Endangered Primate Rescue Center: Report 1996. *Endangered Primate Rescue Center Newsletter* No. 2, 2-12.

Nadler, T. 1998a. The status of Delacour's langur (*Trachypithecus delacouri*) and the possibilities for its long term conservation. Proceedings Workshop on a Conservation Action Plan for the Primates of Vietnam. Hanoi.

Nadler, T. 1998b. Endangered Primate Rescue Center: Report 1997. *Endangered Primate Rescue Center Newsletter* No. 3, 2-10.

Nadler, T. 1998c. Black Langur Rediscovered. *Endangered Primate Rescue Center Newsletter* No. 3.

Nadler, T. 1998d. Wiederentdeckung des Schwarzen Languren (*Trachypithecus francoisi ebenus*). *Zoologische Garten* (N.F.) 68(5): 265-272.

Nadler, T. 1998e. Black Langur Rediscovered. *Asian Primates* 6(3-4): 10-12.

Nadler, T. 1999a. Golden-headed langur (*Trachypithecus poliocephalus*) in the Endangered Primate Rescue Center, Vietnam - and the situation of the species in the wild. *Endangered Primate Rescue Center Newsletter* No. 4:17-19.

Nadler, T. 1999b. Goldschopflangur (*Trachypithecus poliocephalus*) im Endangered Primate Rescue Center, Vietnam und die Bestandssituation in seinem Verbreitungsgebiet. *Zoologische Garten* (N.F.) 69(4): 241-245.

Nadler, T. 2000. Endangered Primate Rescue Center: Report 1999. *Endangered Primate Rescue Center Newsletter* No. 5:1-10.

Nadler, T. and Ha Thang Long 2000. The Cat Ba langur: Past, Present and Future - The Definitive Report on *Trachypithecus poliocephalus*, the World's Rarest Primate. Report Frankfurt Zoological Society, Hanoi.

Nadler, T. and Ha Thang Long 2001. Natural History and Status Review of the Delacour's langur (*Trachypithecus delacouri*). Preliminary Report Frankfurt Zoological Society.

Napier, J. 1967. Evolutionary aspects of primate evolution. *American Journal of Physical Anthropology* 27: 333-342.

Napier, J. R. 1970. Paloecology and catarrhine evolution. In: Napier and Napier (eds.) Old World Monkeys: Ecology, Systematics, and Behaviour. Academic Press, New York.

Napier, J. R., and Napier, P. 1967. A Handbook of Living Primates. Academic Press, London.

Napier, P. 1985. Catalogue of Primates in the British Museum (Natural History) and elsewhere in the British Isles. Part III. Family Cercopithecidae, Subfamily Colobinae. British Museum (Natural History), London.

Newton, P. N., & Dunbar, R. I. M. 1994. Colobine monkey society. In: Davies and Oates (eds.) Colobine Monkeys: their ecology, behaviour and evolution. Cambridge University Press, Cambridge.

Ngo Tu. 1983. Investigation and protection planing. Unpublished report. (Vietnamese).

Ngo Van Tri and Day, A. 1999. The Biological Diversity on the Mammals in Tan Phu State Forest Enterprise. Unpublished report Fauna & Flora International, Indochina-Programme, Hanoi.

Ngo Van Tri and Long, B. 1999. A Report on Survey of Black Gibbon (*Hylobates concolor concolor*) in Son La Province (North Vietnam). Unpublished report Fauna & Flora International, Indochina-Programme, Hanoi.

Ngo Van Tri and Lormée, N. 2000. Survey on primates in Kim Hy Proposed Nature Reserve, Na Ry District, Bac Can Province, Northeastern Vietnam. Unpublished report FFI-Indochina Programme, Hanoi.

Ngo Van Tri. 1999a. Preliminary Assessment on the Mammal in Tuong Limestone Mountain, Ba Thuoc District, Thanh Hoa Province. Unpublished report Fauna & Flora International, Indochina-Programme, Hanoi.

Ngo Van Tri. 1999b. The result on mammals survey in Bien Lac-Nui Ong Nature Reserve, Thanh Linh District, Binh Thuan Province. Unpublished report Fauna & Flora International, Indochina-Programme, Hanoi.

Ngo Van Tri. 2000. Survey of Elephants in Cu Jut and Dak Mil Districts, Dak Lak Province (April 2000*).* Unpublished report Fauna & Flora International, Indochina-Programme, Hanoi.

Ngo Van Tri, 2001. A survey for black gibbons *Nomascus concolor* and other primate species in Van Ban district, Lao Cai province, Northwestern Vietnam. Fauna & Flora International, Indochina-Programme, Hanoi.

Nguyen Ba Thu. 2001. Management and Protection of National Parks and other Protected Areas in Vietnam. Proceedings of the Workshop on environmental education in protected areas of Vietnam: 11-17.

Nguyen Cu. 1990. Preliminary Inventory of *Trachypithecus francoisi poliocephalus* and fauna in Cat Ba National Park. Unpublished report. (Vietnamese).

Nguyen Huu Nhan and Kyung Soo Chun. 1997. Sociocultural Characters of Cat Ba island, Ha Long Bay, Vietnam. In: Ecosystem and Biodiversity of Cat Ba National Park and Halong Bay, Vietnam. Ann. of Nature Conservation, Korean National Cuoncil for Conservation of Nature, vol. 12: Survey of the Natural Environment in Vietnam.

Nguyen Nghia Thin and Jong Won Kim. 1997. The Vegetation of Cat Ba National Park in Vietnam. In: Ecosystem and Biodiversity of Cat Ba National Park and Halong Bay, Vietnam. Ann. of Nature Conservation, Korean National Council for Conservation of Nature, vol. 12: Survey of the Natural Environment in Vietnam

Nguyen Ngoc Chinh, Le Huy Cuong, Nguyen Huy Dung, Vu Van Dung, Nguyen Quoc Dung and Nguyen Huy Thang. 1998. The Phong Nha Caves, Vietnam: World Heritage. List Nomination Form. UNESCO.

Nguyen Phien Ngung. 1998. The status of biodiversity and primate conservation in Cat Ba National Park. Proceedings Workshop on a Conservation Action Plan for the Primates of Vietnam, Hanoi.

Nguyen Xuan Dang, Pham Nhat, Pham Trong Anh and Hendrichsen, D. 1998. Results of survey on mammal fauna in Phong Nha-Ke Bang area, Quang Binh, Vietnam. (A final report of the project - building technical capacity for Phong Nha NR through training and field survey). Report Fauna & Flora International, Indochina-Programme and IEBR, Hanoi.

Nisbett, R. A. and Ciochon, R. L. 1993. Primates in Northern Viet Nam: A review of the Ecology and Conservation Status of Extant Species, with Notes on Pleistocene Localities. *International Journal of Primatology* 14(5): 765-795.

Nooren, H. and Claridge, G. 2001. Wildlife Trade in Laos: the End of the Game. Netherlands Committee for IUCN, Amsterdam.

Nowak, M. 1999. Walker's primates of the world, John Hopkins University Press, Baltimore and London.

Oates, J. F. and Davies, A. G. 1994. What are the colobines? In: Davies and Oates (eds.) Colobine Monkeys: their ecology, behaviour and evolution. Cambridge University Press, Cambridge.

Oates, J. F., Davies, A. G., and Delson, E. 1994. The diversity of living colobines. In: Davies and Oates (eds.) Colobine Monkeys: their ecology, behaviour and evolution. Cambridge University Press, Cambridge.

Oates, J. F. and Trocco, T. F. 1983. Taxonomy and phylogeny of black-and-white colobus monkeys. Inferences from an analysis of loud call variation. *Folia Primatologica* 40: 83-113.

Osgood, W. H. 1932. Mammals of the Kelley-Roosevelts and Delacour Asiatic expeditions. *Field Museum of Natural History, Zoological Series* 18(10): 193-339.

Peng Yanzhang, Ye Zhizhang, Pan Ruliang and Wang Hong. 1993. Observation on the position of genus *Rhinopithecus* in phylogeny. *Primate Report* 35: 63-72.

Peng Yanzhang, Ye Zhizhang, Zhang Yaping and Pan Ruliang. 1989. The classification and phylogeny of snub-nosed monkey (*Rhinopithecus* spp.) based on gross morphological characters. *Zoological Research* 9: 239-248.

Pfeiffer, P. 1969. Considération sur l'Ecologie - Forêts claires du Cambodge Oriental. In: *La Terre et la Vie* no. 1.

Pham Binh Quyen and Truong Quang Hoc. 1997. Study on Socio-economic root causes of biodiversity loss in two distinct regions of Vietnam. Case Studies of Ba Be National Park and Na Hang Nature Reserve in the Mountainous North and Yok Don National Park in the Central Highlands. Final Report. Centre for Natural Resources and Environmental Studies (CRES), Hanoi.

Pham Nhat. 1991. To protect the Tonkin snub-nosed monkey. *Forestry College* 2: 15-17. Xuan May, Vietnam. (Vietnamese).

Pham Nhat. 1992. External characters, Distributions and Status of the leaf eating monkeys in Vietnam. *Forestry College* 1: 34-38. Xuan May, Vietnam. (Vietnamese).

Pham Nhat. 1993. The distribution and status of the douc langur (*Pygathrix nemaeus*) in Vietnam. *Asian Primates* 3(1-2): 2-3.

Pham Nhat. 1994a. Preliminary results on the diet of the Red-shanked douc langur (*Pygathrix nemaeus*). *Asian Primates* 4 (1): 9-11.

Pham Nhat. 1994b. Some data on the food of the Tonkin Snub-nosed Monkey (*Rhinopithecus avunculus*). *Asian Primates* 4 (1): 9-11.

Pham Nhat, Do Quang Huy and Pham Hong Nguyen. 2000. Report on Research Result on Distribution, Ecology and Monitoring Survey of the Red-shanked Douc Langurs *(Pygathrix nemaeus nemaeus)* in Phong Nha-Ke Bang Forest Area. WWF- Indochina.Programme, Hanoi.

Pham Nhat, Do Tuoc and Truong Van La. 1996a. Preliminary survey for the Hatinh Langur in North Central Vietnam. *Asian Primates* 6 (3): 13-17.

Pham Nhat, Do Tuoc and Truong Van La. 1996b. A Survey for Hatinh Langur *Trachypithecus francoisi hatinhensis* in North Central Vietnam. WWF-Indochina Programme, Hanoi and Primate Conservation Inc.

Pham Nhat, Do Tuoc, Tran Quoc Bao, Pham Mong Giao, Vu Ngoc Thanh and Le Xuan Canh. 1998. Distribution and status of Vietnamese primates. Proceedings Workshop on a Conservation Action Plan for the Primates of Vietnam, Hanoi.

Pham Nhat and Nguyen Xuan Dang. 1999. Primates in Phong Nha-Ke Bang area: the status overview and recommendations for further survey and monitoring (Draft). Forestry College of Vietnam, Xuan May, IEBR and WWF-Indochina Programme, Hanoi.

Pham Nhat, Nguyen Xuan Dang and Polet, G. 2001. Field Guide to the Key Mammal Species of Cat Tien National Park. WWF-Cat Tien National Park Conservation Project, Fauna & Flora International, Indochina-Programme and Cat Tien National Park.

Pham Viet Lam and Nguyen Quoc Thang. 1998. Preliminary Study on the Blood of the Black Shanked Doucs housed at the Saigon Zoo and Botanical Garden. Proceedings of Workshop on a Conservation Action Plan for the Primates of Vietnam, Hanoi.

Pham Xuan Xuong. 1997. Investment plan for Kim Hy Nature Reserve, Bac Kan Province. North-western Sub-FIPI, Hanoi. (Vietnamese).

Phung Van Khoa and Lormee, N. 2000. Primate Status Assessment in Bac Kan Province, North Vietnam, January 2000, with a Special Reference to the Black Gibbon *(Hylobates concolor)*. Unpublished report Fauna & Flora International, Indochina-Programme, Hanoi.

Pocock, R. I. 1934. The monkeys of the genera *Pithecus* (or *Presbytis*) and *Pygathrix* found to the east of the Bay of Bengal. *Proceedings of the Zoological Society of London* 1934: 895-961.

Pocock, R. I. 1939. The Fauna of British India, Including Ceylon and Burma: Mammals. 1. Primates and Carnivores (in part), Families Felidae and Viverridae, 2nd ed., Taylor and Francis, London.

Pousarges, E., de 1896. Sur un gibbon d'espèce nouvelle provenant du Haut-Tonkin. *Bulletin du Muséum d'Histoire Naturelle Paris* 2(8): 367-369.

Pousargues, E. 1898. Note préliminaire sur un nouveau Semnopithèque des frontières de Tonkin et de la Chine. *Bull. Mus. d'Hist. Nat.* (1)4: 319-321.

Raffles, T. 1821. Descriptive catalogue of a zoological collection, made on account of the Honourable East India Company, in the Island of Sumatra and its vicinity, under the direction of Sir Thomas Stamford Raffles, Lieutenant Governor of Fort Marlborough: with additional notices of the natural history of those countries. *Trans. Linn. Soc. Lond.* 13(1): 239-276.

Raintree, J. B., Le Thi Phi and Nguyen Van Duong. 1999. Report on a diagnostic survey of conservation problems and development opportunities in the buffer zone of Ke Go Nature Reserve. MARD and Forest-Science Institute of Vietnam.

Ratajszczak, R. 1988. Notes on the Current Status and Conservation of Primates in Vietnam. *Primate Conservation* 9: 134-136.

Ratajszczak, R., Cox, R. and Ha Dinh Duc 1990. A preliminary survey of primates in north Viet Nam. Unpublished Report WWF Project 3869.

Ratajszczak, R., Ngoc Can and Pham Nhat. 1992. A Survey for Tonkin Snub-nosed Monkey (*Rhinopithecus avunculus*) in North Vietnam. FFI Preservation Society, London and WWF International, Gland.

Remane, A. 1921. Beiträge zur Morphologie des Anthropoidengebisses. *Wiegmann-Archiv für Naturgeschichte* 87 (Abt. A)(11): 1-179.

Ren Renmei, Kirkpatrick, R. C., Jablonski, N. G., Bleisch, W. V., and Le Xuan Canh. 1998. Conservation Status and Prospects of the Snub-nosed Langurs (Colobinae: *Rhinopithecus*). In: Jablonski (ed.) The Natural History of the Doucs and Snub-nosed Monkeys. World Scientific Publishing, Singapore.

Richard, A. N. and Russell, L. C. 1993. Primates in North Vietnam: A review of the Ecology and Conservation Status of extant species with notes on Pleistocen localities. *International Journal of Primatology* 14(5): 765-795.

Robinson, H. C. and Kloss, C. B. 1922. New mammals from French Indo-China and Siam. *Ann. Mag. Nat. Hist.* (9th ser.) 9: 87-99.

Robson, C. 1990. Primate surveys in Viet Nam 1989-90, Unpublished report.

Rode, P. 1938. Catalogue des types de mammifères du Museum National d'Histoire Naturelle: Ordre des Primates, Sous-ordre des Simiens. *Bull. Mus. Hist. Nat.* 10, 202-251.

Roos, C. and Nadler, T. 2001. Molecular Evolution of the Douc Langurs. *Zoologischer Garten* (N.F.) 71, 1-6.

Roos, C., Nadler, T., Ya-Ping Zhang and Zischler, H. 2001. Molecular evolution and distribution of the superspecies *Trachypithecus* [*francoisi*]. *Folia Primatol.* 73 (3), 181-182.

Round, P. D. 1999. Avifaunal surveys of the Pu Mat Nature Reserve, Nghe An Province, Vietnam 1998-1999. Report SFNC in Nghe An Province, Vinh, Vietnam.

Rowe, N. 1996. The pictorial guide to the living primates. Pogonias Press, New York.

Rozenddal, F. 1990. Report on surveys in Hoang Lien Son, Lai Chau and Nghe Tinh Provinces. Unpublished report WWF.

Ruempler, U. 1991. Haltung und Zucht von Kleideraffen (*Pygathrix nemaeus nemaeus* Linnaeus 1771) im Koelner Zoo. *Zeitschrift des Koelner Zoo* 34, 47-65.

Ruempler, U. 1998. Husbandry and breeding of Douc langurs *Pygathrix nemaeus* at Cologne Zoo. *Int. Zoo. Yb.* 36, 73-81.

Ruggieri, N. and Timmins, R. J.1995. An initial summary of diurnal primate status in Laos. *Asian Primates* 5 (3): 1-3.

Sarich, V. M., and Cronin, J. E. 1976. Molecular systematics of the primates. In: Goodman, Tashian and Tashian, (eds.) Molecular anthropology. Genes and proteins in the evolutionary ascent of the primates. Plenum Press, New York.

Sawalischin, M. 1911. Der Musculus flexor communis brevis digitorum pedis in der Primatenreihe, mit spezieller Berücksichtigung der menschlichen Varietäten. *Morphologisches Jahrbuch* 42: 557-663.

SFNC Nghe An Province. 2000. Pu Mat: A Biodiversity Survey of a Vietnamese Protected Area. Vinh, Nghe An Province, Vietnam.

Sheeran, L. K. and Mootnick, A. R. 1998. Red-shanked Douc langur *Pygathrix nemaeus*. In: Beacham and Beetz (eds.) Beacham's Guide to International Endangered Species. Beacham Publishing Corp. Osprey, Florida.

Sheeran, L. K. and Mootnick, A. R. 1998. Francois' Langur *Trachypithecus francoisi*. In: Beacham and Beetz (eds.) Beacham's Guide to International Endangered Species. Beacham Publishing Corp. Osprey, Florida.

Sheeran, L. K. and Mootnick, A. R. 1998. Tonkin Snub-nosed monkey *Rhinopithecus avunculus*. In: Beacham and Beetz (eds.) Beacham's Guide to International Endangered Species. Beacham Publishing Corp. Osprey, Florida.

Sibley, C. G. and Ahlquist, J. E. 1984. The phylogeny of hominoid primates, as indicated by DNA-DNA hybridization. *Journal of Molecular Evolution* 20: 2-15.

Sibley, C. G. and Ahlquist, J. E. 1987. DNA hybridization evidence of hominoid phylogeny: Results from an expanded data set. *Journal of Molecular Evolution* 26: 99-121.

Simonetta, A. 1957. Catalogo e sinonimia annotata degli ominoidi fossili ed attuali (1758-1955). *Atti Soc. Toscana Sci. Nat., Pisa* Ser. B 64: 53-113.

Srivastava, A., Biswas, J., Das, J. and Bujarbarua, P. 2001. Status and Distribtion of Golden langurs (*Trachypithecus geei*) in Assam, India. *Am. Journ. of Primatology* 55: 15-23.

Stanford, C. B. 1988. Ecology of the capped langur and Phayre's leaf monkey in Bangladesh. *Primate Conservation* 9: 125-128.

Stenke, R. 2001a. Artenschutzprojekt Goldkopflangur auf Cat Ba Island, Nordvietnam. *Mitt. Zool. Soc. for the Conservation of Species and Populations* 17(1): 6-8.

Stenke, R. 2001b. Golden-headed langur, where are you? *IPPL-News* 28(2): 13-14.

Stenke, R. 2001c. Unpublished ZSCSP-report to Conservation International.

Stern, K. and Stern, J. 1967. Im Urwald. *Das Magazin* (7): 4 pages.

Supriatna, J., Maullang, B. O. and Soekara, E. 1986. Group composition, home range and diet of the maroon leaf monkey (*Pesbytis rubicunda*) at Tanjung Putting Reserve, Central Kalimantan, Indonesia. *Primates* 27: 185-190.

Tan Bangjie. 1955. Apes in China. *Bull. of Biology* 3, 17-23. (Chinese).

Tan Bangjie. 1957. Rare catches by Chinese animal collectors. *Zoo Life* 12(2): 61-63.

Tan Bangjie. 1985. The status of primates in China. *Primate Conservation* 5: 63-81.

Thomas, O. 1921. A new monkey and a squirrel from the Middle Mekong, on the eastern frontier of Siam. *Ann. Mag. Nat. Hist.* (9)7: 181-183.

Thomas, O. 1928. The Delacour exploration of French Indo-China. – Mammals II: Mammals collected during the winter of 1926-1927. *Proceedings of the Zoological Society, London* 1928(1): 139-150.

Thomas, O. 1929. The Delacour exploration of French Indo-China. – Mammals III: Mammals collected during the winter of 1927-1928. *Proceedings of the Zoological Society, London* 1928(4): 831-841.

Thompson, S. D. 1998. Demographic and Genetic Evaluation - Francois langur ssp. population. In Pate, D. E. (ed.) Francois' Langur *Studbook*.

Timmins, R.J. and Evans, T.D. 1996. A Wildlife and Habitat survey of Nakai-Nam Theun National Biodiversity Conservation Area, Khammouan and Bolikhamsai Provinces, Lao PDR. Wildlife Conservation Society, Vientiane.

Timmins, R. J. and Khounboline, K. 1996. A preliminary wildlife and habitat survey of Hin Namno National Biodiversity Conservation Area, Khammouane Province, Lao PDR. CPAWM and WCS, Vientiane.

Timmins, R. J. and Soriyun, M. 1998. A wildlife survey of the Tonle San and Tonle Srepok river basins in Northeastern Cambodia. Fauna & Flora International, Indochina-Programme, Hanoi and Wildlife Protection Office, Phnom Penh.

Timmins, R. J., Do Tuoc, Trinh Viet Cuong and Hendrichsen, D. K. 1999. A Preliminary Assessment of the Conservation Importance and Conservation Priorities of the Phong Nha-Ke Bang National Park, Quang Binh Province, Vietnam. Fauna & Flora International, Indochina-Programme, Hanoi.

Timmins, R. J. and Duckworth, J. W. 1999. Status and Conservation of Douc Langurs (*Pygathrix nemaeus*) in Laos. *Int. Journal of Primat.* 20, 469-489.

Tordoff, A. W., Siurua, H., and Sobey, R. 1997. Ben En National Park, Biodiversity Survey 1997. Frontier Vietnam Forest Research Programme (Technical Report No. 12), Hanoi and London.

Tordoff, A. W., Swan, S., Grindley, M., and Siurua H. 1999. Hoang Lien Nature Reserve, Biodiversity survey and conservation evaluation 1997/98. Frontier Vietnam Forest Research Programme (Technical Report No. 13), Hanoi and London.

Tordoff, A. W., Tran Hieu Minh and Tran Quang Ngoc. 2000a. A Feasibility Study for the Establishment of Ngoc Linh Nature Reserve, Quang Nam Province, Vietnam. BirdLife International, Vietnam programme and FIPI (Conservation Report No. 10), Hanoi.

Tordoff, A. W., Tran Quang Ngoc, Le Van Cham and Dang Thang Long. 2000b. A rapid Field Survey of Five Sites in Bac Kan, Cao Bang and Quang Ninh Provinces, Vietnam. A Review of the Northern Indochina Subtropical Forests Ecoregion. BirdLife International Vietnam Programme and FIPI (Conservation Report Number 14), Hanoi.

Tordoff, A. W., Le Trong Dat and Hardcastle, J. 2001. A Rapid Biodiversity Survey of Che Tao Commune, Mu Cang Chai District, Yen Bai Province, Vietnam. Technical report to the Danida funded project: Improved conservation through institutional strengthening in Cambodia, Lao PDR and Vietnam. BirdLife International, Vietnam Programme and Fauna & Flora International, Indochina-Programme, Hanoi.

Traitel, D. A. 1996. Building a safe haven for Douc Langurs. *Zoonooz* 69(10): 10-15.

Tran Sa. 1998. The Human-Hearted Monkey. *Vietnam Cultural Window* 9: 14-15.

Trang Ngoc But. 1995. Biological Rescources of Cat Ba National Park. Proceedings of The National Conference on National Parks and Protectes Areas of Vietnam, Hanoi.

Trinh Dinh Hoang. 2001. Report on surveys for eastern black crested gibbon (*Nomascus* sp. cf. *nasutus*) and other primates in Hoa An and Tra Linh districts, Cao Bang province. Unpublished report Fauna & Flora International, Indochina-Programme, Hanoi.

Trinh Viet Cuong. 2000. Present status of Elephants (*Elephas maximus*) in Kon Plong District (Kon Tum Province) Unpublished report Fauna & Flora International, Indochina-Programme, Hanoi.

Trinh Viet Cuong and Ngo Van Tri. 2000. Survey of Elephants in Cu Jut District, Dak Lak Province and Kon Plong District, Kon Tum Province (December 1999 to January 2000*).* Unpublished report Fauna & Flora International, Indochina-Programme, Hanoi.

Trouessart, E. L. 1911. On a new species of Semnopithecus (*Semnopithecus poliocephalus*) from Tonkin. *Ann. Mag. Nat. Hist.* (8th ser.) 8: 271-272.

UNDP. 1997. Creating Protected Areas for Resource Conservation using Landscape Ecology (PARC). VIE/95/G31, VIE/95/G31/1G/31 and VIE/95/G41/A/1G/99.

U.S. Board on Geographic Names. 1986. Gazetteer of Vietnam, Vols. I and II, Defense Mapping Agency, Washington DC.

Van Peenen, P. F. D., Ryan, P. F. and Light, R. H. 1969. Preliminary Identification Manual for Mammals of South Vietnam. United States National Museum, Smithsonian Institution, Washington DC.

Van Schaik, C. P., Assink, P. R. and Salafsky, N. 1992. Territorial behaviour in Southeast Asian langurs: Resource defence or mate defence? *American Journal of Primatology* 26: 233-242.

Vermeulen, J. and Whitten, T. 1999. Biodiversity and Cultural Property in the Management of Limestone Resources. The World Bank, Washington DC.

VRTC. 1997. Biological inventory survey in Vu Quang Nature Reserve Huong Khe District, Ha Tinh Province, Vietnam (July-September 1997). Draft report WWF-Vu Quang Conservation Project by participants of the VRTC expedition.

Vu Ngoc Long. 2001. Biodiversity and Social Studies of Nui Chua Nature Reserve, Ninh Thuan Province, Vietnam. Institute of Tropical Biology, Sub-Institute of Ecology, Resources and Environmental Studies, Ho Chi Minh City.

Walston, J. and Vinton, M. 1999. A Wildlife and Habitat Survey of Hin Namno National Biodiversity Conservation area and Adjacent Areas, Khammouane Province, Lao PDR. (Biodiversity Survey Report 1) WWF Lao, Vientiane.

Wang, W., Forstner, M. R. J., Zhang, Y. P., Liu, Z. M., Wei, Y., Huang, H. Q., Hu, H. G., Xie, Y. X., Wu, D. H. and Melnick, D. J. 1997. Phylogeny of Chinese leaf monkeys using mitochondrial ND3-ND4 gene sequences. *International Journal of Primatolology* 18(3): 305-320.

Wang, W., Su ,B., Lan, H., Zhang, Y.P., Lin, S. Y., Liu, R. Q., Liu, A. H., Hu, H. G., Xie, Y. X. and Wu, D. H. 1995. rDNA difference and phylogenetic relationship of two species of golden monkeys and three species of leaf monkeys. In: Xia Wuping and Zhang Yongzu (eds.). Primate Research and Conservation. China Forestry Publishing House, Beijing.

Wang Yingxiang, Jiang Xuelong and Feng Qing 1998. Taxonomy, status and conservation of leaf monkeys in China. Proc. Int. Seminar on Commemorating the Naming of Francois' leaf Monkey for its First centenary and Protection the Primate. Wuzhou, China.

Wege, D. C., Long, A. J., Mai Ky Vinh, Vu Van Dung and Eames, J. C. 1999. Expanding the protected areas network in Vietnam for the 21st century. An analysis of the current system with recommendations for equitable expansion. BirdLife International, Vietnam Programme and FIPI (Conservation Report No. 6), Hanoi.

Weitzel, V. 1992a. A Review of the Taxonomy of *Trachypithecus francoisi. Australian Primatology* 7(2): 2-4.

Weitzel, V. 1992b. Proposal: Field Survey of *Trachypithecus francoisi* in Northern Vietnam; Winter 1992-93 and Associated Travel. Submitted for Consideration to the Royal Zoological Society of South Australia.

Weitzel, V. and Groves, C. P. 1985. Nomenclature and taxonomy of the colobine monkeys of Java. *International Journal of Primatolology* 6(4): 397-409.

Weitzel, V. and Vu Ngoc Thanh. 1992. Taxonomy and conservation of *Trachypithecus francoisi* in Vietnam. *Asian Primates* 2(2): 2-5.

Weitzel, V., Yang, C. M. and Groves, C. P. (1988). A catalogue of primates in the Singapore Zoological Reference Collection. *The Raffles Bulletin of Zoology* 36: 1-166.

Wienberg, J., and Stanyon, R. 1987. Fluorescent heterochromatin staining in primate chromosomes. *Human Evolution* 2: 445-457.

Wikramanayake, E. D., Vu Van Dung and Pham Mong Giao. 1997. A Biological and Socio-economic Survey of West Quang Nam Province with Recommendations for a Nature Reserve. UNDP, WWF and MARD, Hanoi.

Wirth, R. 1998. Tonkin snub-nosed (*Rhinopithecus avunculus*) rediscovered. *Asian Primates* 2(2): 1-2.

Wirth, R., Adler, H. J. and Nguyen Quoc Thang. 1991. Douc Langurs: How Many Species Are There? *Zoonooz* 64(6): 12-13.

Wislocki, G. B. 1929. On the placentation of primates, with a consideration of the phylogeny of the placenta. *Contributions to Embryology* 20(111): 51-80.

Wislocki, G. B. 1932. On the female reproductive tract of the gorilla, with a comparison of that of other primates. *Contributions to Embryology* 23(135): 163-204.

World Bank. 1995. Vietnam-Environmental Programmes and Policy Priorities for a Socialist Economy in Transition.

Wu, M. 1993. The present status of primates in Guangxi. *Chinese Primate Research and Conservation News* 2(1): 7-8.

Xu, L., Liu, Z. and Yu, S. 1983. Mammalia. In: Xu, Liu and Yi (eds.) Birds and mammals of Hainan Island. Scientific Publishing Agency, Beijing. (Chinese).

Zehr, S., Ruvolo, M., Heider, J. and Mootnick, A. 1996. Gibbon phylogeny inferred from mitochondrial DNA sequences. *American Journal of Physical Anthropology* Suppl. 22: 251 (Abstract).

Zhang Yaping and Ryder, O. A. 1998. Mitochondrial Cytochrome b Gene sequences of Langurs: Evolutionary Inference and Conservation Relevance. In: Jablonski (ed.) The Natural History of the Doucs and Snub-nosed Monkeys. World Scientific Publishing, Singapore.

Zhang Yongzu, Quan Guoqiang, Zhao Tigong and Southwick, C. H. 1992. Distribution of primates (except *Macaca*) in China. *Acta Theriologica Sinica* 12: 81-95.

Zheng Shuyi. 1997. Report on the golden monkey research and conservation workshop. *Asian Primates* 6(3-4): 20.

Zhaoyuan Li, Zimin Liu, Yi Wei and Rogers, E. Decline in Population Density of the White-headed Langur, *Presbytis lı ucocephalus*, in China (in print).

Zimmermann, E. 1990. Differentiation of vocalizations in bushbabies (Galaginae, Prosimiae, Primates) and the significance for assessing phylogenetic relationships. *Zeitschrift für Zoologische Systematik und Evolutionsforschung* 28: 217-239.

Appendix 1

Leaf monkey species in the protected area network of Vietnam
(The list follows an arragnement of the protected areas from north to south)

Protected area	Province	Trachypithecus francoisi	T. laotum hatinhensis	T. l. hatinhensis var. ebenus	T. p. poliocephalus	T. delacouri	T. crepusculus	T. germaini	Pygathrix nemaeus	P. cinerea	P. nigripes	Rhinopithecus avunculus
Muong Nhe NR	Lai Chau						C					
Du Gia NR	Ha Giang	P										C
Tay Con Linh I NR												E
Kim Hy NR	Bac Kan	C										
Ba Be NP		C										P
Na Hang NR	Tuyen Quang	C										C
Cham Chu NR	Tuyen Quang/ Ha Giang	P										C
Phuong Hoang-Thanh Xa CHS	Thai Nguyen	P										
Cat Ba NP	Hai Phong				C							
Huong Son CHS	Ha Tay					C						
Cuc Phuong NP	Ninh Binh/ Hoa Binh/ Thanh Hoa					C	E					
Ngoc Son NR	Hoa Binh					C						
Thuong Tien NR							U					
Hoa Lu CHS	Ninh Binh					C						
Van Long NR						C						
Pu Luong NR	Thanh Hoa					C	P					
Xuan Lien NR							P					
Ben En NP							C					
Pu Hoat NR	Nghe An						P					
Pu Huong NR							P					
Pu Mat NP							C		C			
Vu Quang NP	Ha Tinh						P		C			
Ke Go NR									P			
Phong Nha- Ke Bang NP	Quang Binh		C	C					C			
Khe Net NR			P									
Dak Rong NR	Quang Tri								P			
Phong Dien NR	Thua Thien Hue								P			
Bach Ma NP									C			

Protected area	Province	*Trachypithecus francoisi*	*T. laotum hatinhensis*	*T. l. hatinhensis var. ebenus*	*T. p. poliocephalus*	*T. delacouri*	*T. crepusculus*	*T. gemaini*	*Pygathrix nemaeus*	*P. cinerea*	*P. nigripes*	*Rhinopithecus avunculus*
Bana- Nui Chua NR	Da Nang								P			
Ban Dao Son Tra NR									C			
Song Thanh NR	Quang Nam								P	C		
Ba To CHS	Quang Ngai									C		
Mom Ray NP	Kom Tum							U	P?	P?	P	
Chu Prong NR	Gia Lai							P				
Kon Cha Rang NR									C	P?		
Kon Ka Kinh NR									C	P?		
A Yun Pa NR											P?	
Yok Don NP	Dak Lak							C			P	
Chu Hoa NR									P?	P?	P?	
Chu Yang Sin NR								U			C	
Nam Nung NR											P	
Bi Dup- Nui Ba NR	Lam Dong							U			C	
Deo Ngoan Muc NR											P	
Cat Tien NP	Lam Dong/ Dong Nai/ Binh Phuoc							C			C	
Bu Gia Map NR	Binh Phuoc										C	
Nui Chua NR	Ninh Thuan							P			C	
Bien Lac- Nui Ong NR	Binh Thuan							P			C	
Lo Go Sa Mat NR	Tay Ninh							P				
Nui Ba Den CHS											U	
Hon Chong CHS	Kien Giang							C				
Kien Luong NR								C				

C	Occurrence confirmed
P	Provisional occurrence
P?	Provisional occurrence: *Pygarthrix* species not identifyed
U	Unknown
E	Probably extinct
NP	National park
NR	Nature reserve
CHS	Cultural and historical site

Appendix 2

Selected areas of high value for primate conservation

The following list provides further details about several locations cited in this work. The localities have been chosen among others because they are representative of habitat, human impact and conservation status in different regions of Vietnam

The forest type classification follows Wege *et al.* (1999).

For each locality the leaf-monkey and gibbon species recorded in the area are listed. For provisional occurrence the species is noted in brackets.

Muong Nhe Nature Reserve (LAI CHAU)
Special use forest: Nature Reserve (314,000 ha)
Forest size: about 47,000 ha, very fragmented
Forest type: lowland evergreen, montane evergreen (in majority secondary), mixed, bamboo
Elevation: ~400m to 2,124m a.s.l. (Mount Phu Nam Man)
Leaf monkey and gibbon species: *Trachypithecus crepusculus*, [*Nomascus leucogenys leucogenys*]

Muong Nhe Nature Reserve was established in 1986, and covers about 182,000 ha. It was extended in March 1997 and the current area totals 314,000 ha. (BirdLife International & FIPI, 2001). Even though it is the largest protected area in Vietnam, 250,000 ha are actually agricultural land and only 15 % of the nature reserve is covered by forest. A major part of the reserve (66 %) is covered by grassland, and the vast majority of the remaining forest is secondary.

The topography is dominated by mountains of medium height. The average height of these mountains is around 1,200m a.s.l. but several peaks are above 1,800m a.s.l., and the highest point, Mount Phu Nam Man, reaches 2,124m a.s.l. The population has increased very rapidly during recent years. 10,000 to 12,000 immigrants, mostly H'mong, moved into this area between 1989 and 1997. Shifting cultivation is predominant and the forested areas are still decreasing (28% of the reserve in 1991; Cox et al., 1992 : versus under 20% in 1997; Hill et al., 1997). One of the biggest threats to the fauna is hunting. Cox et al. (1992) estimated that there were more than 200 elephants in the early 1970s. Information from local people indicates that elephants had largely been eradicated from the area by 1990, and that, by 2000, gaur and tiger had met with the same fate. No conservation interventions are implemented in this nature reserve. Located in a sensitive border area with Lao PDR, access for field surveys is very difficult.

Than Xa Forest, Vo Nhai District (THAI NGUYEN)
Special use forest: Partly included in Phuong Hoang-Than Xa Cultural and Historical Site
Forest size: NA
Forest type: limestone
Elevation: NA
Leaf monkey and gibbon species: [*Trachypithecus francoisi*], [*Nomascus nasutus*]

During their survey in March 1998, Geissmann & Vu Ngoc Thanh (2000) repeatedly met hunters and loggers in Than Xa forest, and hunting pressure appeared to be particularly high in this area. The resident human population of Than Xa is relatively high (about 2,000 people). Furthermore, Than Xa harbours an even larger population of immigrant gold miners (about 10,000 people) who add to hunting pressure in the forest. Geissmann & Vu Ngoc Thanh (2000) reported that every morning, about 300 people climbed the path leading from La Hien to Kim Son Communes in order to transport food to gold miners working in the "mining city" adjacent to Xuyen Son.

Inhabitants of Xuyen Son, who had traditionally collected their drinking water from a little river which they had channelled in picturesque canals running through the whole village, are no longer able to do so, because the mining city was erected upstream of the village. Because the chemicals used to bind the gold poisoned the river, the villagers now have to collect and carry their drinking water from water holes and brooks situated further up the mountain (Geissmann & Vu Ngoc Thanh, 2000).

Ba Be National Park and proposed extension (BAC KAN)
Special use forest: Ba Be National park (7,610 ha) and proposed extension (15,730 ha)
Forest size: about 13,000 ha (including extension)
Forest type: limestone and lowland evergreen
Elevation: 178m (Ba Be lake) to 1,098m a.s.l.
Leaf monkey and gibbon species: *Trachypithecus francoisi*, [*Rhinopithecus avunculus*]

Ba Be was decreed a cultural and historical site in 1977 and established as a national park in 1992. The strictly protected area covers 7,610 ha including 300 ha of lake surface (BirdLife International & FIPI, 2001). A feasibility study to extend the area of the park to 23,340 ha, with a buffer zone of 9,538 ha, was conducted in 1995 (BirdLife International & FIPI, 2001).

Ba Be is a complicated system of lakes, rivers, streams, and limestone mountains. Most of the forest is secondary, having been exploited by humans for a long time. A total of 2,871 people from the Tay, H'mong and Kinh ethnic groups live inside the national park (BirdLife International & FIPI, 2001). Levels of disturbance are generally high, and selective logging and clearance for agriculture are commonplace. Consequently, much of the forest in the national park is disturbed and few areas of undisturbed forest remain (Hill et al., 1997). Non-timber forest products are still collected. Signs of poaching were regularly recorded during a survey in 1996 (Hill et al., 1996a).

Cuc Phuong National Park (NINH BINH, HOA BINH, THANH HOA)
Special use forest: National park (22,000 ha)
Forest size: 20,479 ha
Forest type: limestone
Elevation: 100m to 636m a.s.l.
Leaf monkey and gibbon species: Trachypithecus delacouri, [Trachypithecus crepusculus]

Cuc Phuong was the first protected area to be established in Vietnam, and was decreed in 1962. In 1966 it was upgraded to the first Vietnamese national park, currently with an area of 22,000 ha.

The national park lies at the south-eastern end of a limestone range that runs north-west to Son La Province. This limestone is predominantly karst, and the range rises sharply out of the surrounding plains, to elevations of up to 636m a.s.l.

Cuc Phuong has an extremly rich flora. The national park is considered to be one of seven globally significant centres of plant diversity in Vietnam (Davis *et al.*, 1995). Cuc Phuong supports populations of several mammal species of conservation importance, including the critically endangered Delacour's langur.

Six hamlets with 650 people were relocated from the central valley and two villages from the Buoi river valley. However, there are still 2,000 people living inside the national park. The buffer zone and the surrounding area is home to around 100,000 people, many of whom use natural resources of the national park. The most widely exploited forest products are timber and fuelwood. The collection of snails, mushrooms, bamboo shoots, banana stems and medicinal plants are common. Hunting, both for subsistence and commercial purposes, takes place at an unsustainable level, and threatens to eradicate a number of mammal, bird and reptile species from the national park.

The large number of tourists who visit Cuc Phuong pose particular problems for the management. Waste disposal, collection of plants, and excessive noise created by large tour groups are problems which are not under control. More significantly, the management agenda of the national park is heavily focused on tourism development at the expense of biodiversity conservation. This has resulted in the development of tourism infrastructure with negative environmental impacts. Upgrading the road through the central valley of the national park has facilitated exploitation of forest products, and the construction of artificial lakes inside the national park has resulted in forest clearance, altered local hydrology, and introduction of exotic fish species.

Currently, one of the biggest threats to biodiversity is the planned construction of National Highway No. 2 through the national park. If the planned road development goes ahead, it will bisect the national park. Apart from the direct impact of construction, the road would facilitate access to the forest and, hence, forest product extraction, and might, in the future, act as a focus for human settlements.

Xuan Lien Nature Reserve (THANH HOA)
Special use forest: Nature reserve (21,000 ha)
Forest size: 18,522 ha (Le Trong Trai, 1999a)
Forest type: lowland evergreen, lower montane mixed coniferous, broadleaf evergreen, mixed bamboo
Elevation: reaching 1,605m a.s.l.
Leaf monkey and gibbon species: [*Trachypithecus crepusculus*], *Nomascus leucogenys leucogenys*

Xuan Lien Proposed Nature Reserve near the Lao border was decreed in 1999 and covers 21,000 ha. The topography is characterised by medium-high mountains which reach elevations of 800m to 1,600m a.s.l., and are dissected by deep, narrow valleys. It includes 18,522 ha of forest, only 1,572 ha of which is primary. A large proportion comprises bamboo forest (36.8%) with varying degrees of disturbance. Undisturbed forest only occurs above 700m a.s.l. The primary forest is now distinctly fragmented (Le Trong Trai, 1999a).

Thanh Hoa Provincial FPD (2000) identifies the biggest threats to biodiversity to be forest clearance for agriculture, hunting, over-exploitation of forest products and forest fire. Since the establishment of the nature reserve in 2000 considerable patrolling, enforcement and awareness activities by highly motivated park management and rangers have contributed to more effective conservation.

Pu Luong Nature Reserve (THANH HOA)
Special use forest: Nature reserve (17,662 ha)
Forest size: 13,305 ha (BirdLife International & FIPI, 2001)
Forest type: lowland evergreen, lower montane evergreen, limestone
Elevation: 60m to 1,667m a.s.l.
Leaf monkey and gibbon species: *Trachypithecus delacouri*; [*Trachypithecus crepusculus*], [*Nomascus leucogenys leucogenys*]

Pu Luong was officially established in 1999, and covers 17,662 ha (BirdLife International & FIPI, 2001). The nature reserve lies along two parallel mountain ridges, that run from north-west to south-east, and is bisected by a central valley. This valley contains several human settlements and a large area of agricultural land, and, hence, is not included within the nature reserve. The two mountain ridges have starkly contrasting landforms based on their different substrates. The smaller, south-western one is made up of mostly igneous and metamorphic rock, and consists of rounded forested hills and wide, shallow valleys. The larger, north-eastern ridge is composed of heavily dissected limestone karst, and is a continuation of the limestone range that runs from Cuc Phuong National Park to Son La Province.

It is likely that Pu Luong has strong faunal and floral affinities with Cuc Phuong National Park. However, the higher elevation at Pu Luong and the presence of more extensive areas of montane evergreen forest means that Pu Luong can be expected to support a number of species that do not occur at Cuc Phuong. Consequently, the fauna and flora of the two sites are complementary, and the conservation of both sites is necessary to conserve the full range of biodiversity of the limestone range (BirdLife International & FIPI, 2001).

Ben En National Park (THANH HOA)
Special use forest: National Park (38,153 ha)
Forest size: NA, heavily fragmented
Forest type: lowland evergreen, limestone (patches), mixed bamboo
Elevation: 20m to 497m a.s.l.
Leaf monkey and gibbon species: *Trachypithecus crepusculus, Nomascus leucogenys leucogenys*

Ben En was designated a nature reserve in 1986 and was upgraded to a national park in 1992. The core zone covered 16,634 ha. An extension incresed the area of the core zone to 38,153 ha. Unfortunately people have already moved into the extension area and cleared forest for swidden agriculture and sugar cane cultivation (BirdLife International & FIPI, 2001).

Four state forest enterprises conducted commercial logging activities until 1992 (Tordoff et al., 2000). No area in Ben En remains undisturbed. As a result of the commercial logging operations the forest is characterised by small, shade-intolerant trees and a dense undergrowth dominated by bamboo.

The population sizes of most large mammal species are very low, probably as a result of past hunting, and some have undoubtedly been eradicated already (Tordoff et al., 2000). More than 10,000 people inhabit the buffer zone and still use forest products such as bamboo shoots, rattan, timber and fuel wood.

Pu Hoat Nature Reserve (NGHE AN)
Special use forest: Nature reserve (67,934 ha)
Forest size: 33,555 ha (lightly disturbed forest)
Forest type: lowland evergreen, lower montane evergreen, upper montane evergreen
Elevation: 100m to 2,452m a.s.l. (Mount Pu Hoat)
Leaf monkey and gibbon species: [*Trachypithecus crepusculus*], *Nomascus leucogenys leucogenys*

The nature reserve was established in 1997 to cover 67,934 ha, of which 56,837 ha were said to be under strict protection (BirdLife International & FIPI, 2001). The nature reserve lies along the ridge of mountains that form the border between Vietnam and Laos PDR, and adjacent to Xuan Lien Nature Reserve in Thanh Hoa Province. The majority of the site lies between 800m and 1,400m a.s.l. The highest point is Mount Pu Hoat at 2,452m a.s.l. The topography of the nature reserve is characterised by high, rugged mountain ridges, interspersed with steep-sided valleys.

Forests in the southern part of the nature reserve are highly disturbed, mainly due to agricultural encroachment by H'mong communities. In the higher elevation of the northern part, near Xuan Lien Nature Reserve, the natural habitat is better preserved (Le Trong Trai, pers. comm.). Forest loss has been indentified as one of the biggest threats to biodiversity, as habitat fragmentation leads to species loss. The second major threat is hunting, which has already led to rapid decline in large and medium sized mammals, and the local extinction of some globally threatened species (Osborn *et al.*, 2000).

Interview data indicate that elephant and tiger occurred at the nature reserve until the mid 1980s but are now extinct in the area. Despite designation no nature reserve management board has been established. No conservation activities are currently implemented.

Pu Huong Nature Reserve (NGHE AN)
Special use forest: Nature Reserve (49,845 ha)
Forest size: less than 20,000 ha (Kemp et al., 1997)
Forest type: lowland evergreen, montane evergreen, (both with small deciduous elements)
Elevation: 200m to 1,560m a.s.l. (Mount Phu Lon)
Leaf monkey and gibbon species: [*Trachypithecus crepusculus*], *Nomascus leucogenys leucogenys*

Pu Huong Nature Reserve was decreed in 1986 to comprise 5,000 ha. In 1995 the Nhe An Provincial People's Committee approved the establishment of a nature reserve which covers 49,845 ha. The nature reserve lies at the northern extent of the Annamite mountain range, separated by the Ca river.

The topography of the nature reserve is steep and mountainous, and is dominated by ridges of mountains, of 950m to 1,560m a.s.l.

Lowland evergreen forest is heavily disturbed and dominated by member of diptoerocarps, while areas that were previously subjected to commercial timber extraction have now regenerated into mature forest. In the lower montane forest disturbance is restricted to more accessible areas at lower elevations.

An important part of Pu Huong forest remains undisturbed. But forest clearance for agriculture is significant. Each year, as much as 200 ha are cleared inside the boundaries (Kemp & Dilger, 1996). Illegal logging is also intensive. Hunting represents a major threat to biodiversity, particularly to any populations of large mammal species that remain. Species of conservation concern, such as Saola and White-cheeked gibbon, were regularly hunted (Kemp & Dilger, 1996), and may be extirpated.

Pu Mat National Park (NGHE AN)
Special use forest: National park (91,113 ha)
Forest size: NA
Forest type: lowland evergreen, lower montane evergreen
Elevation: 100m to 1,841m a.s.l. (90% of the national park is under 1,000m a.s.l.)
Leaf monkey and gibbon species: *Trachypithecus crepusculus*, *Pygathrix nemaeus*, *Nomascus leucogenys siki*

Two protected areas established in 1986 were combined to provide the legal basis for establishing Pu Mat Nature Reserve in 1995 with an area of 91,113 ha. The nature reserve was upgraded to national park status in 2001.

The area is made up of steep mountains up to 1,841m a.s.l., and numerous, usually steep river valleys in between the mountains. The steep terrain in most parts of the national park has been an obstacle to extensive clearance of forest. Where the river valleys are broader, there is some cultivation, and three villages are situated inside the national park. Forest is virtually gone in the main valleys and heavily disturbed on the hills adjacent to the main valleys and villages. Approximately 50% of the forest in the core zone is undisturbed, 20% is lightly disturbed and 30% is heavily disturbed (Geissmann *et al.*, 2000).

As Pu Mat supports populations of most probably eight primate species, several of global conservation concern, it should be considered a high priority area for primate conservation in Vietnam. While these primate species still occur in Pu Mat, their density is very low. Hunting threatens the population of many mammal and bird species. Other threats to biodiversity include agricultural enchroachment and gold mining.

Vu Quang National Park (HA TINH)
Special use forest: National park (55,029 ha)
Forest size: 38,300 ha (BirdLife International & FIPI, 2001).
Forest type: lowland evergreen, lower montane evergreen, medium montane evergree, upper montane evergreen, elfin
Elevation: 30m to 2,286m a.s.l. (Mount Rao Co)
Leaf monkey and gibbon species: *Pygathrix nemaeus*, [*Trachypithecus crepusculus*], *Nomascus leucogenys siki*

Vu Quang was protected during the French colonial period and comprised an area of about about 30,000 ha. Because of this classification the area was closed to local communities. In the 1960s a forest enterprise was established. The main management objective was forest exploitation; commercial logging activities continued until 1993 (Eve, 2000).

In 1986 a nature reserve was established comprising an area of 16,000 ha. In 1994 an investment plan was approved which proposed establishing a 55,950 ha nature reserve, and in 2002 this was upgraded to a national park.

At low elevations, the landscape is largely anthropogenic and consists of many elements including human habitation, agricultural land, grassland and scrub. Between 100m and 500m a.s.l., much of the forest has been selectively logged, although, above 500m a.s.l. the slopes are covered by primary forest (BirdLife International & FIPI, 2001).

In 1992, the attention of the world scientific community was focussed on Vu Quang Nature Reserve, following the discovery of a large mammal species, Saola *Pseudoryx nghetinhensis* (Vu Van Dung *et al.*, 1993).

A variety of human activities threaten the biodiversity of Vu Quang Nature Reserve. Clearance of forest for agriculture and the development of human settlements is destroying natural habitats and fragmenting forest cover. Hunting levels are intensive and nearly every species of mammal and bird is potential prey for hunters. Illegal timber extraction takes place and logging trails can be found everywhere. The quantity of firewood removed from the nature reserve each year is equivalent to the clear-felling of about 428 ha of forest (BirdLife International & FIPI, 2001). Cattle can be found throughout the nature reserve at all times of year and the extraction of fragrant oil from several tree species is an extremely destructive process.

A further threat is the planned development of National Highway No. 2, the proposed route of which cuts through the nature reserve. If the road development is to go ahead, it will have an irreversible impact on the nature reserve. Besides fragmenting habitat and facilitating access to the forest, road construction may lead to human settlement close to the core areas of the nature reserve (Eve, 2000).

Ke Go Nature Reserve (HA TINH)
Special use forest: Nature reserve (24,801 ha)
Forest size: 24,284 ha (Le Trong Trai et al., 1996)
Forest type: lowland evergreen
Elevation: mainly below 300m, reaching 500m a.s.l. (Moc Buoi)
Leaf monkey and gibbon species: [*Pygathrix nemaeus*], *Nomascus leucogenys siki*

Ke Go Nature Reserve was established in 1996, and covers 24,801 ha. The topography is comprised of gently undulating low hills, a landscape typical for the midlands of central Vietnam (BirdLife International & FIPI, 2001). The nature reserve supports 24,284 ha of natural forest, equivalent to 98% of the total area. However, the forest has been selectively logged in the past; 76% is classified as heavily disturbed, and undisturbed primary forest is virtually absent.

About 40,000 people live in the buffer zone. A large proportion of the population is dependant on forest products collected in the nature reserve. The major threats to biodiversity are hunting, illegal timber extraction, charcoal production, fuelwood, rattan and medicinal plant collection and fragrant oil extraction (Le Trong Trai *et al.*, 1999).

Ke Go is of important conservation value because it comprises one of the largest remaining lowland evergreen forests in Central Vietnam (Le Trong Trai et al., 1996). Several threatened and endemic species inhabit the nature reserve and it supports populations of five restricted-range birds including the only known population of Vietnamese pheasant, *Lophura hatinhensis*, in the world.

Phong Nha-Ke Bang National Park (Quang Binh)
Special use forest: National park (85,754 ha)
Forest size: NA
Forest type: limestone, lowland evergreen, riparian (142 ha)
Elevation: reaching 400m a.s.l.
Leaf monkey and gibbon species: *Trachypithecus laotum hatinhensis*, *Trachypithecus laotum hatinhensis* var. *ebenus*, *Pygathrix nemaeus*, *Nomascus leucogenys siki*

A 5,000 ha cultural and historical site was established in 1986. In 1993 an investment plan was approved to change the status to a nature reserve and to extend the area to 41,132 ha. A revised investment plan incorporated Ke Bang limestone area, and in 2001 the nature reserve was upgraded to a national park with a total of 85,754 ha.

The national park is located in one of the largest areas of contiguous limestone in Indochina, which also includes Hin Namno National Biodiversity Conservation Area in Laos PDR. The topography of the national park is characterised by precipitous karst ridges, which rise to elevations of around 400m a.s.l. Scattered among these ridges are narrow valleys and pockets of igneous rock formations (BirdLife International & FIPI, 2001).

The nature of the terrain has restricted human encroachment into limestone areas (Timmins *et al.*, 1999). As a result the limestone karst is almost entirely forested, apart from steep cliff faces. The forest has been cleared in flat valleys within the limestone massif, and in lowland areas bordering it.

The limestone forest ecosystem supports a high diversity of plant and animal species, including a number endemic to the karst massif. Of great conservation significance are several species found at the site that are endemic to this part of central Vietnam and Laos PDR. The national park is an important site for primate conservation. Nine primates species occur in the area including the Hatinh langur which is endemic to this karst complex.

Currently the biggest threat to biodiversity is hunting. Most hunting is commercially oriented, with a well established wild animal trade in the area. There has been substantial commercial hunting of primates, which has resulted in major population declines.

Illegal timber trade is highly organised, and it is not unusual to witness the extraction of up to 1,000 kg of timber a day (BirdLife International & FIPI, 2001). Extraction is focussed on economically valuable timber, and these species are becoming increasingly rare.

Tourism development is another threat to biodiversity at the site. Already the Quang Binh Tourism Company is attempting to both promote the national park's natural assets and manage the burgeoning numbers of visitors. Ecotourism studies by WWF have warned of the potential risk of uncontrolled tourism development. There is, as yet, no adequate provision for ecotourism at the national park (BirdLife International & FIPI, 2001).

Dakrong and Phong Dien Nature Reserves (QUANG TRI and THUA THIEN-HUE)
Special use forest: Nature reserves (82,074 ha)
Forest size: 53,721 ha (Le Trong Trai & Richardson, 1999a)
Forest type: lowland evergreen
Elevation: most below 500m, reaching 1,615m a.s.l. (Co Pung in Phong Dien Nature Reserve)
Leaf monkey and gibbon species: [*Pygathrix nemaeus*], *Nomascus leucogenys siki or gabriellae*

The Dakrong Watershed Protection Forest was upgraded in 2000 to a nature reserve comprising 40,526 ha.

The Phong Dien Nature Reserve in the neighbouring province is contiguous with Dakrong Nature Reserve. An investment plan for Phong Dien Nature Reserve to comprise 41,548 ha was approved in 2001(BirdLife International & FIPI, 2001).

The topography of both reserves is dominated by a ridge of low mountains, which extends south-east from the Annamite mountains. The nature reserves support the largest remaining area of lowland evergreen forest in central Vietnam. However, the forest at Dakrong has been heavily disturbed, and primary forest only occupies around 60% of the total area.

The impact of war has been dramatic to the areas although, whilst the indirect legacy of war continues to exert an influence on habitats and wildlife, new threats are now more significant.

About 90,000 people inhabit the buffer zone (Le Trong Trai & Richardson, 1999a). The main threats to biodiversity are hunting (particularly through use of snares), collection of firewood and other non-timber forest products, illegal logging, forest fires (caused by swidden cultivation, the deliberate setting of fires to collect metal from bomb and shell casings, and spontaneous detonation of unexploded ordnance), and clearance of forest land for agriculture (BirdLife International & FIPI, 2001).

Bach Ma National Park (THUA THIEN-HUE)
Special use forest: National Park (22,031 ha)
Forest size: 39,089 ha (Anon. 1997d)
Forest type: lowland evergreen, montane evergreen, bamboo
Elevation: reach 1,448m a.s.l. (Mount Bach Ma)
Leaf monkey and gibbon species: *Pygathrix nemaeus*, [*Nomascus leucogenys siki or gabriellae*]

In 1925, a proposal was made to create a 50,000 ha national park in the Hai Van region, primarily in order to protect Edward's Pheasant *Lophura edwardsi*. Subsequently, several areas of what is now Bach Ma National Park were designated forest reserves. In 1986, the Bach Ma-Hai Van National Park was decreed with a total area of 40,000 ha, and in 1990 was split into the current Bach Ma National Park comprising 22,031 ha and Bac Hai Van and Nam Hai Van Cultural and Historical Site (BirdLife International & FIPI, 2001).

The national park lies on a high mountain ridge that runs west-east from the Laotian border to the East Sea at the Hai Van pass. This ridge interrupts the coastal plain of Vietnam, and, therefore, forms a biogeographic boundary between the faunas and floras of northern and southern Vietnam. The national park is dominated by a west-east ridge with several peaks above 1,000m a.s.l. This ridge also affects the local climate, which is probably the wettest in Vietnam: the mean annual rainfall at the summit of Mount Bach Ma is 7,977 mm (WWF/EC, 1997). The terrain is steep, with slopes averaging 15-250.

The national park has long been noted for its rich biological diversity. One reason is that, within a relatively small area, it supports a wide range of habitat types, from coastal lagoon to montane forest. Additionally, Bach Ma is situated at a biographic boundary between northern and southern Vietnam (BirdLife International & FIPI, 2001).

Most of the park was subject to defoliant spraying during the American war. After the reunification, timber companies worked in the park. Logging was officially stopped in 1989. However, illegal exploitation and forest products collection, such as rattan and fuelwood, still continue. The result is that large areas of the park are now deforested and no forest remains undisturbed.

The development of a third Vietnamese economic stage in Hue-Da Nang area should attract new migrants and increase the human pressure on the park. The introduction of exotic tree species for reforestation programmes in the buffer zone constitues another threat for the integrity of Bach Ma (Anon, 1995d).

Ban Dao Son Tra Nature Reserve (DA NANG)
Special use forest: Nature reserve (4,439 ha)
Forest size: 1,635 ha (BirdLife International & FIPI, 2001).
Forest type: lowland evergreen, heavily degraded
Elevation: reaching 696m a.s.l. (Son Tra)
Leaf monkey and gibbon species: *Pygathrix nemaeus* (*most probably eradicated*)

The nature reserve was established in 1977 on a peninsula marking the eastern boundary of Da Nang Bay. The protected area covers about 4,400 ha (Anon. 1989). The forest was heavily damaged by the defoliant spraying during the American war. The only patches of less disturbed forest occur on the northern slopes between 300m and 650m a.s.l.

The remnant forest is of poor quality, and the population of Red-shanked douc langur has been reduced to a very small number, and has probably already been eradicated.

Song Thanh Nature Reserve and Nam Giang District (QUANG NAM)
Special use forest: Partly included in Song Thanh Nature Reserve (93,249 ha)
Forest size: 88,879 ha inside nature reserve
Forest type: lowland evergreen, montane evergreen
Elevation: NA, numerous peaks over 1,000m a.s.l.
Leaf monkey and gibbon species: *Pygathrix cinerea*, [*Pygathrix nemaeus*]

Song Thanh(-Dak Pring) Nature Reserve covers much of the area which was established in 1999. The topography of the area is mountainous, with numerous peaks over 1,000m a.s.l. Song Thanh Nature

Reserve is contiguous with Ngoc Linh Nature Reserve (Kon Tum). It is, therefore, part of one of the largest areas of contiguous conservation coverage in Vietnam.

About 64,000 people inhabit three districts which rely heavily on forest land and resources for their livelihood. Therefore, shifting cultivation, fuelwood collection, illegal logging and wildlife trapping are major threats to biodiversity (Le Nho Nam, 2001).

Mom Ray National Park (KON TUM)
Special use forest: Nature reserve (56,621 ha)
Forest size: NA
Forest type: lowland evergreen, lower montane evergreen, lowland semi-deciduous (smaller areas)
Elevation: 200m to 1,773m a.s.l. (Mount Chu Mom Ray)
Leaf monkey and gibbon species: [*Pygathrix sp.*], [*Pygathrix nigripes*], [*Trachypithecus germaini*], [*Nomascus gabriellae*]

Mom Ray was decreed a nature reserve in 1982 covering 10,000 ha. It was extended to 48,658 ha in 1996, and upgraded to national park in 2002. The national park is situated in an area of medium-high mountains, the highest of which is Mount Chu Mom Ray at 1,773m a.s.l.

The national park may be one of the best remaining areas for tiger in Vietnam, with a population estimated at 10 to 15 individuals in 1997 (Duckworth & Hedges, 1998).

The forest area is an important source of forest products for local communities, who experience, on average, two months of food shortage per year. Main threats for biodiversity include shifting cultivation, fuelwood, bamboo and rattan collection and hunting.

Kon Plong District (KON TUM)
Special use forest: None
Forest size: NA
Forest type: semi-deciduous, montane evergreen
Elevation: reaching 2,266m a.s.l. (Ngoc Kring)
Leaf monkey and gibbon species: [*Pygathrix nigripes*], [*Pygathrix cinerea*]

The topography of Kon Plong is primarily hilly, and high mountains are located in the west part adjoining Dac To District. The terrain makes the forest very difficult to access and almost all the forests of the district are situated on such terrain (Trinh Viet Cuong & Ngo Van Tri, 2000).

At present, human settlements and cultivation affect the forests in the center and along Lo River and the eastern part of the district adjoining Quang Ngai Province. Forest has been converted into agricultural land in many places (Trinh Viet Cuong & Ngo Van Tri, 2000).

In general, mammal fauna is still abundant. Hunting tools such as guns and snares are frequently seen and openly used, but traps, bows and arrows are less common.

Kon Ha Nung area (GIA LAI)
Special use forest: Partly included in Kong Cha Rang (15,900 ha) and Kon Ka Kinh (41,710 ha) Nature Reserves
Forest size: 15,610 ha (Kon Cha Rang); 33,565 (Kon Ka Kinh)
Forest type: lowland evergreen, lower montane evergreen, Kon Ka Kinh: mixed coniferous and broadleaf

Elevation: reaching 1,452m a.s.l. (Mount Kong Cha Rang); 570m to 1,748m a.s.l. (Mount Kon Ka Kinh)
Leaf monkey and gibbon species: [*Trachypithecus germaini*], *Pygathrix nemaeus*, [*Pygathrix cinerea*], [*Pygathrix nigripes*], *Nomascus gabriellae*

In the area, two nature reserves, Kong Cha Rang and Kon Ka Kinh were established in 1986, covering 15,900 and 41,710 ha respectively (BirdLife International & FIPI, 2001).

A mountainous plateau dominates the topography of the nature reserve. A number of summits attain altitudes greater than 1,000m a.s.l. in the northern part. The highest point is Mount Kon Cha Rang at 1,452 m a.s.l. The lowest point at the site is about 800m a.s.l.

Until 1975, people lived within the boundaries of what is now Kon Cha Rang Nature Reserve, although, today, there are no people living inside the nature reserve. The population of the buffer zone is about 6,000. The four main threats to biodiversity at Kon Cha Rang are perceived to be forest clearance for plantations, hunting, exploitation of forest products and forest fire.

Kon Ka Kinh Nature Reserve supports a range of montane habitat types across an altitudinal range from 700m to 1,748m a.s.l. To the north of the Kon Ka Kinh Nature Reserve is Mount Ngoc Linh, the highest mountain in the Central Highlands. To the south and west, the topography is flatter, and altitudes are below 500m a.s.l. There are several mountain peaks above 1,500m a.s.l.

Of particular importance are 2,000 ha of mixed coniferous and broadleaf forest, and within the protected areas system of Vietnam, this vegetation sub-type is known only from this area. 20% of the nature reserve has been degraded by past commercial logging activities and continuing illegal timber extraction. A further 29% has been cleared by commercial logging or shifting cultivation and now supports a range of secondary vegetation types.

Because of the low population density and large area of unused, fertile land, the buffer zone has been, and continues to be, a focus for spontaneous migration from other parts of Vietnam. Spontaneous migration is one of the most serious conservation problems at Kon Ka Kinh, because, as the population of the buffer zones increases, so does pressure on forest resources (Le Trong Trai *et al.* 2000).

In Kon Ka Kinh, most of illegal logging is done by Kinh people from northern Vietnam travelling in the area for this purpose. Rattan collection is intensive and this non-timber forest product was reported to be becoming increasingly rare in the forest (Le Trong Trai, 2000).

Kong Cha Rang and Kon Ka Kinh Nature Reserves are linked by an area of forest currently under the management of Dak Roong and Tram Lap Forest Enterprises. The boundaries of the protected areas should be revised to make them contiguous (Le Trong Trai, 2000).

Yok Don National Park and proposed extension (DAK LAK)
Special use forest: National park (58,000 ha) and proposed extension (57,545 ha)
Forest size: 54,200 ha (Le Xuan Canh et al., 1997a)
Forest type: deciduous, semi-deciduous, lowland evergreen
Elevation: mostly 200m, reaching 482m a.s.l. (Mount Yok Don)
Leaf monkey and gibbon species: *Trachypithecus germaini*, [*Pygathrix nigripes*]

In the 1986 decree, Yok Don was listed as Tien Seo Ea Sup Nature Reserve. It was upgraded to a national park in 1991. It covers 58,200 ha and an extension of 57,345 ha.

The topography of most of the site is flat, at an elevation of around 200m a.s.l. Remote sensing data indicate that the majority of the forest at Yok Don is deciduous forest, with smaller areas of semi-deciduous and evergreen forest. Yok Don has a reputation as an important site for the conservation of large mammals. The occurance of elephant, gaur, banteng and tiger is confirmed (Duckworth & Hedges, 1998), but hunting is the greatest direct threat to the fauna. Surveys demonstrated that, despite the availibility of significant areas of suitable habitat, population sizes of key mammal and bird species have declined over recent years, indicating that current management actions are not stabilising populations of this species (BirdLife International & FIPI, 2001).

Migration into the area surrounding Yok Don National Park is leading to an increase in human pressure on natural ressources, and threatens to undermine conservation activities at the national park. Dak Lak Province has the highest immigration rate of any province in Vietnam. In a five-year period (1990 to 1995) the population of the province rose by 21%. The increased demand for forest products, including timber and wild animals, contributes to biodiversity loss at Yok Don National Park.

Cu Jut District (DAK LAK)
Special use forest: Partly included (2,782 ha) in proposed extension of Yok Don National Park
Forest size: NA
Forest type: deciduous, semi-deciduous, lowland evergreen forest
Elevation: 300m to 500m a.s.l.
Leaf monkey and gibbon species: *Trachypithecus germaini*; *Pygathrix nigripes*; [*Pygathrix cinerea*]; *Nomascus gabriellae*

Cu Jut is located on hills with several large streams running into Srepok river. Deciduous and semi-deciduous forests are the main type of vegetation. Evergreen forest is found only in the central part of the district (Trinh Viet Cuong & Ngo Van Tri, 2000).
In 1992, the human population was approximately 22,500 people. Following the relocation of ethnic groups from provinces in the north, the population increased to about 88,000 people in 1997 accoss the whole district (Trinh Viet Cuong & Ngo Van Tri, 2000).
The forest fringes are subject to cultivation and industrial use. The eastern part of the district is a densely populated resettled area and newcomers continue to arrive. It seems probable that the flat land, agricultural soils and increased population pressures will promote increased rates of deforestation in the future (Trinh Viet Cuong & Ngo Van Tri, 2000).

Da Lat Plateau (DAK LAK, LAM DONG, KHANH HOA and NINH THUAN)
Special use forest: Partly included in Chu Yang Sinh (59,278 ha), Bi Dup-Nui Ba (71,062 ha) and Deo Ngoan Muc (2,000 ha) Nature Reserves
Forest size: NA
Forest type: Bi Dup-Nui Ba Nature Reserve: coniferous, evergreen, broadleaf evergreen (small parts)
 Chu Yang Sinh Nature Reserve: semi-decidiuos, lowland evergreen, montane evergreen, elfin
 Deo Ngoan Muc Nature Reserve: evergreen, lower montane evergreen (heavily degraded)
Elevation: Bi Dup-Nui Ba Nature Reserve: above 1,400m reaching 2,287m a.s.l. (Mount Bi Dup),
 Chu Yang Sinh Nature Reserve: 300m to 2,442m a.s.l.
 Deo Ngoan Muc Nature Resreve: 200m to 980m a.s.l.
Leaf monkey and gibbon species: [*Trachypithecus germaini*], *Pygathrix nigripes*, [*Nomascus gabriellae*]

The area is comprised of a mountainous plateau forming the southern edge of the western highlands and includes three nature reserves. The change in elevation is large.

Shifting cultivation, practiced by ethnic minorities has had a great impact on vegetation composition. This activity has reduced the area of evergreen forest and led to its fragmentation. Most remaining areas of evergreen forest are on mountains and steep slopes. Frequent use of fire prevents the regeneration of evergreen forest and promotes the development of a vegetation dominated by *Pinus kesiya*, which is now the most widespread forest type on the plateau (est. 120,000 ha in 1990). Furthermore, lands covered by evergreen forest are preferred by shifting cultivators, because of a better soil quality. Evergreen forest is of a higher biodiversity value than pine forest. Consequently, shifting cultivation is a major threat to biodiversity. However, *Pinus kesiya* is a pioneer species and evergreen species grow predominantly in the understory. Unfortunately, much of the pine forest understorey on the Plateau is burnt annually (Eames, 1995).

Logging is a major activity on the Da Lat Plateau. As the principle market for charcoal and fuelwood is Da Lat city, the continued expansion of this urban centre is likely to lead to increased pressure on the forest resources.

Other non-timber forest products collected include pine resin, extracted from *Pinus kesiya*, oils collected from *Cinnamomum* species, and flowering plants, in particular members of the Orchidaceae (Eames, 1995).

The spontaneous arrival of new migrants, and governmental programmes to develop new economic zones, increase pressure on the environment (Eames, 1995).

Chu Yang Sin Nature Reserve was established by governmental decree in 1986, and comprises about 20,000 ha, in 1993 extended to 34,237 ha, and in 1997 to 59,278 ha (BirdLife International & FIPI, 2001). The forest comprises 33,827 ha.

In the frame of the EU funded project "*Expanding the Protected Areas Network in Vietnam for the 21st Century*" an extension to the southwest and to the east (in Khanh Hoa Province) of Bi Dup-Nui Ba Nature Reserve propose to add 45,600 ha of land. A 16,400 ha extension to south of Chu Yang Sin Nature Reserve (in Lam Dong Province) would connect this protected area with Bi Dup-Nui Ba. This complex of protected areas should cover 168,737 ha and is considered a priority for forest conservation (Wege *et al.*,1999).

Bi Dup-Nui Ba Nature Reserve was established in 1993 on an area which comprises two separate earlier- established small nature reserves. The reserve supports high levels of plant diversity and endemism. The fauna is also very species rich, and exhibits high levels of endemism.

The overall level of human impact on the nature reserve is moderate. One of the greatest threats comes from shifting cultivation. Other main causes of forest loss have been charcoal production and fuelwood collection. Charcoal production has already lead to the destruction of most of the evergreen forest on Mount Lang Bian.

Primates are subject to hunting on the Da Lat Plateau. An unquantified trade of wildlife trophies occurs in Da Lat and Bao Loc cities, principally for sale to local tourists. In addition, significant numbers of primates are sent to Ho Chi Minh City markets (Eames & Robson, 1993).

Cat Tien National Park (LAM DONG, DONG NAI, BINH PHUOC)
Special use forest: National Park (73,878 ha)
Forest size: NA
Forest type: primary and secondary lowland evergreen, semi-deciduous, flooded, mixed bamboo
Elevation: about 100m to 659m a.s.l.
Leaf monkey and gibbon species: *Pygathrix nigripes, Trachypithecus germaini, Nomascus gabriellae*

The first decision degreed the establishment of a nature reserve in 1978 comprising 35,000 ha. Following the rediscovery of the Javan Rhinoceros (*Rhinoceros sondaicus*) in 1989, a Rhinoceros

Sanctuary of 30,635 ha, close to the nature reserve, was established in 1992. Following a decision in 1998 the two areas were combined, forming a total of 73,878 ha.

The topography of the national park varies greatly between the three sectors. The Cat Loc sector is situated at the western end of the Central Highlands and, is consequently, rather hilly. The Nam Cat Tien and the Tay Cat Tien sectors are situated in the lowlands of southern Vietnam. The topography is characterised by low rolling hills, the highest of which reaches an elevation of 372m a.s.l.

Cat Tien National Park supports a variety of habitat types, including primary and secondary forest, flooded forest, grassland and areas dominated by bamboo. The national park is one of the most important sites for the conservation of large mammals in Vietnam. The most significant one is that of Javan Rhinoceros. This is the only known population of the subspecies *R. s. annamiticus* in the world (BirdLife International & FIPI, 2001).

There are four main conservation challanges at the national park: forest land is being converted into agricultural land; illegal exploitation of timber, rattan, mammals, birds and fish is common; land within the national park designated as agricultural land is not under the management of the park management board, and poor coordination among different local government agencies is resulting in an emphasis on agricultural development instead of biodiversity conservation; and a plan exists to construct two hydro-electric dams, which would reduce the size of important wetlands and have implications for migratory and resident waterbird species, fish species and grazing mammals (G. Polet In: BirdLife International & FIPI, 2001).

Appendix 3

IUCN RED LIST CATEGORIES AND CRITERIA

Since their inception, the IUCN Red Data Books and Red Lists have enjoyed an increasingly prominent role in guiding conservation activities of governments, NGO's and scientific institutions. The IUCN Red Lists are widely recognized as the most comprehensive, apolitical global approach for evaluating the conservation status of plant and animal species. The introduction in 1994 of a scientifically rigorous approach to determine risks of extinction which is applicable to all species and infra-specific taxa, has become a virtual world standard (Hilton-Taylor, 2000).

The goals of the IUCN Red List Programme are to:

* Provide a global index of the state of degeneration of biodiversity; and

* Identify and document those species most in need of conservation attention if global extinction rates are to be reduced.

For the 1996 Red List, the risk of extinction was evaluated for all known species of birds and mammals. For the 2000 Red List the primates were the only mammals to be comprehensively reassessed. Much of the taxonomic revision by Groves (2001) has been adopted by the IUCN-Species Survival Commission -Primate Specialist Group- and used in the 2000 Red List. Some results on recent investigations, especially based on DNA are also considered in this Primate Status Review.

The name *IUCN Red List of Threatened Species* implies that the primary focus is at the species level, the 2000 Red List also includes all assessments that have been made at the infra-specific and sub-populations level.

An analysis and identification of the countries with the largest number of threatened species enables countries to be informed about their global responsibility to protect and maintain the biodiversity for which they are ultimately the stewards.

The results of the mammals are shown in Figures 1 and 2. The primates make up a considerable portion of the listing for Vietnam.

Figure 1.

The twenty countries with the largest numbers of threatened mammal species (Hilton-Taylor, 2000).

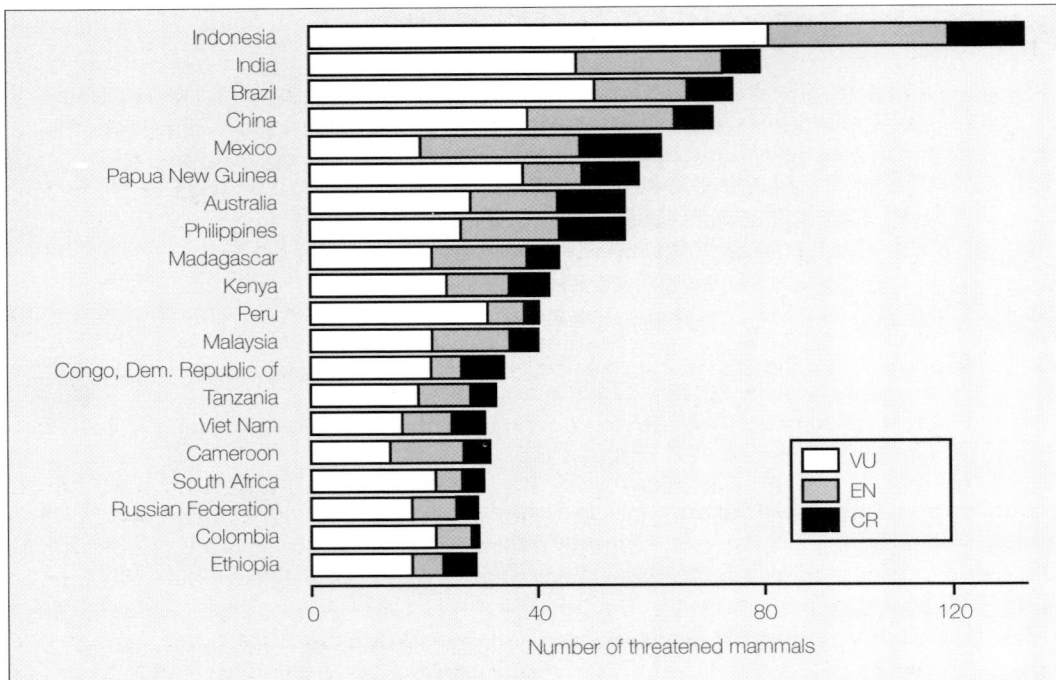

Figure 2.

The twenty countries with the largest numbers of threatened mammal species, with their numbers as a percentage of total mammal diversity found in each country (Hilton-Taylor, 2000).

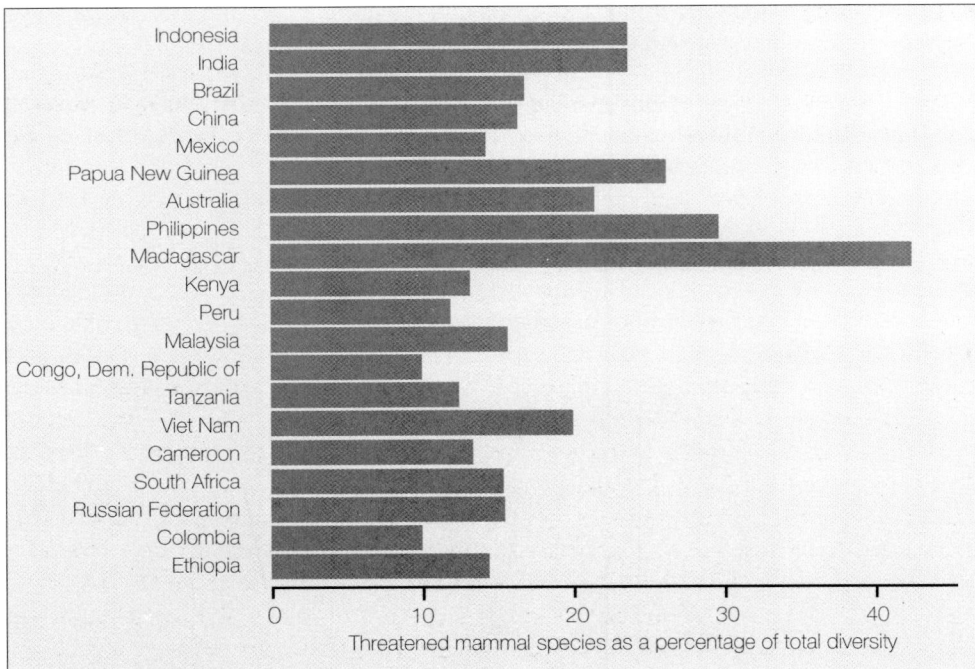

EXTINCT (EX)

A taxon is Extinct when there is no reasonable doubt that the last individual has died.

EXTINCT IN THE WILD (EW)

A taxon is Extinct in the wild when it is known only to survive in cultivation, in captivity or as a naturalized population (or populations) well outside the past range. A taxon is presumed extinct in the wild when exhaustive surveys in known and/or expected habitat, at appropriate times (diurnal, seasonal, annual), throughout its historic range have failed to record an individual. Surveys should be over a time frame appropriate to the taxon's life cycle and life form.

CRITICALLY ENDANGERD (CR)

A taxon is Critically Endangered when it is facing an extremly high risk of extinction in the wild in the immediate future, as defined by any of the criteria (A to E).

ENDANGERED (EN)

A taxon is Endangered when it is not Critically Endangered but is facing a very high risk of extinction in the wild in the near future, as defined by any of the criteria (A to E).

VULNERABLE (VU)

A taxon is Vulnerable when it is not Critically Endangered or Endangered but is facing a high risk of extinction in the wild in the medium-term future, as defined by any of the criteria (A to E).

LOWER RISK (LR)

A taxon is Lower Risk when it has been evaluated, does not satisfy the criteria for any of the categories Critically Endangered, Endangered or Vulnerable. Taxa included in the Lower Risk category can be separated into three subcategories:

* Conservation Dependent (cd). Taxa which are the focus of a continuing taxon-specific or habitat-specific conservation programme targeted towards the taxon in question, the cessation of which would result in the taxon qualifying for one of the threatened categories above within a period of five years.
* Near Threatened (nt). Taxa which do not qualify for Conservation Dependent, but which are close to qualifying for Vulnerable.
* Least Concern (lc). Taxa which do not qualify for Conservation Dependent or Near Threatened.

DATA DEFICIENT (DD)

A taxon is Data Deficient when there is inadequate information to make a direct or indirect assessment of its risk of extinction based on its distribution and/or population status. A taxon in this category may be well studied, and its biology well known, but appropriate data on abundance and/or distribution is lacking. Data Deficient is therefore not a category of threat or Lower Risk. Listing of taxa in this category indicates that more information is required and acknowledges the possibility that future research will show that threatened classification is appropriate. It is important to make positive use of whatever data are available. In many cases great care should be exercised in choosing between DD and threatened status. If the range of a taxon is suspected to be relatively circumscribed, if a considerable period of time has elapsed since the last record of the taxon, threatened status may well be justified.

NOT EVALUATED (NE)

A taxon is Not Evaluated when it is has not yet been assessed against the criteria.

Figure 3.
Structure of the IUCN Red List Categories

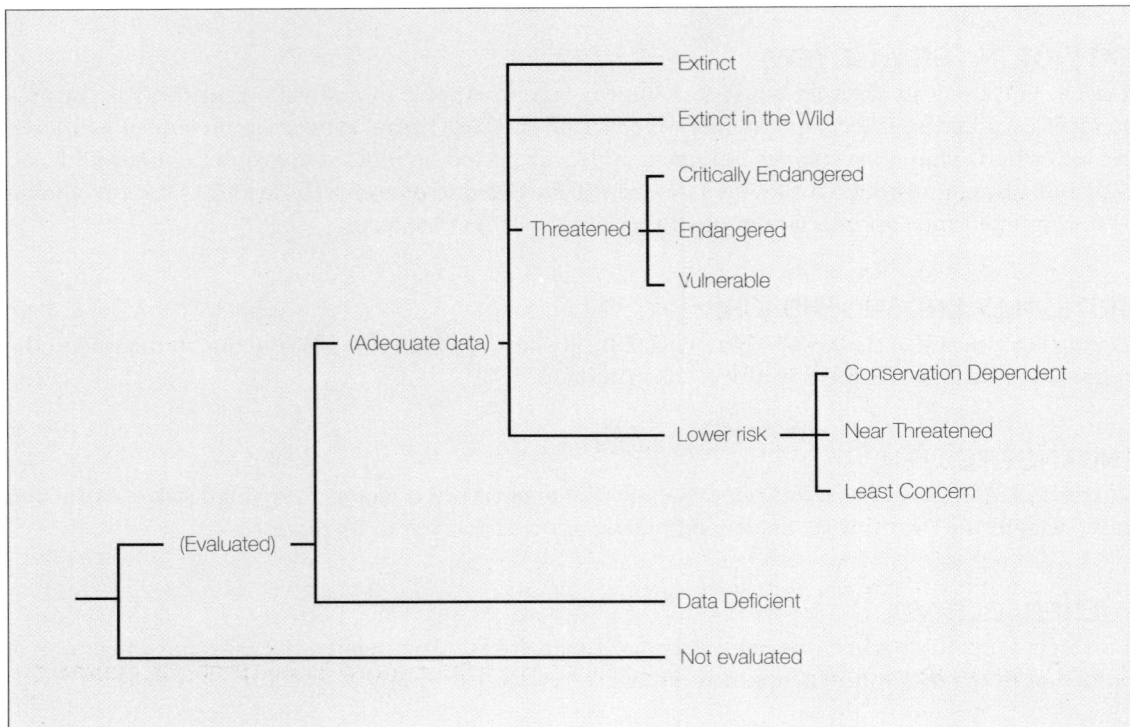

The IUCN Red List Criteria
for CRITICALLY ENDANGERED - ENDANGERED - VULNERABLE

CRITICALLY ENDANGERED (CR)

A   Population reduction in the form of either of the following:

1   An observed, estimated, inferred or suspected reduction of at least 80% over the last ten years or three generations, whichever is the longer, based on (and specifying) any of the following:

   a   direct observation
   b   an index of abundance appropriate for the taxon
   c   a decline in area of occupancy, extent of occurrence and/or quality of habitat
   d   actual or potential level of exploitation
   e   the effects of introduced taxa, hybridisation, pathogens, pollutants, competitors or parasites

2   A reduction of at least 80%, projected or suspected to be met within the next ten years or three generations, whichever is the longer, based on (and specifying) any of b, c, d or e above.

B   Extent of occurrence estimated to be less than 100km^2 or area of occupancy estimated to be less than 10km^2 , and estimates indicating any two of the following:

1   Severely fragmented or known to exist at only a single location

2   Continuing decline, observed, inferred or projected, in any of the following:

    a   extent of occurrence
    b   area of occupancy
    c   area, extent and/or quality of habitat
    d   number of locations or subpopulations
    e   number of mature individuals

3   Extreme fluctuations in any of the following:

    a   extent of occurrence
    b   area of ocupancy
    c   number of locations or subpopulations
    d   number of mature individuals

C   Population estimated to number less than 250 mature individuals and either:

1   An estimated continuing decline of at least 25% within three years or one generation, whichever is longer or

2   A continuing decline, observed, projected, or inferred in numbers of mature individuals and population structure in the form of either:

    a   severely fragmented (i.e. no subpopulation estimated to contain more than 50 mature individuals)
    b   all individuals in a single subpopulation

D   Population estimated to number less than 50 mature individuals.

E   Quantitative analysis showing the probability of extinction in the wild is at least 50% within ten years or three generations, whichever is longer.

ENDANGERED (EN)

A **Population reduction in the form of either of the following:**

1   An observed, estimated, inferred or suspected reduction of at least 50% over the last ten years or three generations, whichever is longer, based on (and specifying) any of the following:

    a   direct observation

    b   an index of abundance appropriate for the taxon

    c   a decline in area of occupancy, extent of occurrence and/or quality of habitat

    d   actual or potential level of exploitation

    e   the effects of introduced taxa, hybridisation, pathogens, pollutants, competitors or parasites

2   A reduction of at least 50%, projected or suspected to be met within the next ten years or three generations, whichever is longer, based on (and specifying) any of b, c, d or e above.

B **Extent of occurrence estimated to be less than 5,000km^2 or area of occupnacy estimated to be less than 500km^2, and estimates indicating any two of the following:**

1   Severely fragmented or known to exist at no more than five locations.

2   Continuing decline, observed, inferred or projected, in any of the following:

    a   extent of occurrence

    b   area of ocupancy

    c   area, extent and/or quality of habitat

    d   number of locations or subpopulations

    e   number of mature individuals

3   Extreme fluctuations in any of the following.

    a   extent of occurrence

    b   area of ocupancy

    c   number of locations or subpopulations

    d   number of mature individuals

C **Population estimated to number less than 2500 mature individuals and either:**

1   An estimated continuing decline of at least 20% within five years or two generations, whichever is longer or

2   A continuing decline, observed, projected, or inferred in numbers of mature individuals and population structure in the form of either:

    a   severely fragmented (i.e. no subpopulation estimated to contain more than 250 mature individuals)

    b   all individuals in a single subpopulation

D   Population estimated to number less than 250 mature individuals.

E   Quantitative analysis showing the probability of extinction in the wild is at least 20% within 20 years or five generations, whichever is longer.

VULNERABLE (VU)

A   Population reduction in the form of either of the following:

1   An observed, estimated, inferred or suspected reduction of at least 20% over the last ten years or three generations, whichever is longer, based on (and specifying) any of the following:

    a   direct observation
    b   an index of abundance appropriate for the taxon
    c   a decline in area of occupancy, extent of occurrence and/or quality of habitat
    d   actual or potential level of exploitation
    e   the effects of introduced taxa, hybridisation, pathogens, pollutants, competitors or parasites

2   A reduction of at least 20%, projected or suspected to be met within the next ten years or three generations, whichever is longer, based on (and specifying) any of b, c, d or e above.

B   Extent of occurrence estimated to be less than 20,000km^2 or area of occupancy estimated to be less than 2,000km^2, and estimates indicating any two of the following:

1   Severely fragmented or known to exist at no more than ten locations.

2   Continuing decline, observed, inferred or projected, in any of the following:

    a   extent of occurrence
    b   area of occupancy
    c   area, extent and/or quality of habitat
    d   number of locations or subpopulations
    e   number of mature individuals

3   Extreme fluctuations in any of the following:

    a   extent of occurrence
    b   area of occupancy
    c   number of locations or subpopulations
    d   number of mature individuals

C   **Population estimated to number less than 10,000 mature individuals and either:**

1   An estimated continuing decline of at least 10% within ten years or three generations, whichever is longer or

2   A continuing decline, observed, projected, or inferred in numbers of mature individuals and population structure in the form of either:

    a   severely fragmented (i.e. no subpopulation estimated to contain more than 1,000 mature individuals)

    b   all individuals in a single subpopulation

D   **Population very small or restricted in the form of either of the following:**

1   Population estimated to number less than 1,000 mature individuals

2   Population is characterized by an acute restriction in its area of occupancy (typically less than 100km²) or in the number of locations (typically less than five). Such a taxon would thus be prone to the effects of human activities (or stochastic events whose impact is increased by human activities) within a very short period of time in an unforeseeable future, and is thus capable of becoming Critically Endangered or even Extinct in a very short period

E   **Quantitative analysis showing the probability of extinction in the wild is at least 10% within 100 years.**